Meeting on Nonequilibrium Thermodynamics and Statistical Physics: From Rational Modeling to Its Applications 2017

Journal of Physics: Conference Series Volume 1035

Fukuoka, Japan
16 - 17 March 2017

Editors:

Shigeru Taniguchi
Takashi Arima

ISBN: 978-1-5108-6410-8
ISSN: 1742-6588

Printed from e-media with permission by:

Curran Associates, Inc.
57 Morehouse Lane
Red Hook, NY 12571

Some format issues inherent in the e-media version may also appear in this print version.

Printed by Curran Associates, Inc. (2018)

For permission requests, please contact the Institute of Physics
at the address below.

Institute of Physics
Dirac House, Temple Back
Bristol BS1 6BE UK

Phone: 44 1 17 929 7481
Fax: 44 1 17 920 0979

techtracking@iop.org

Additional copies of this publication are available from:

Curran Associates, Inc.
57 Morehouse Lane
Red Hook, NY 12571 USA
Phone: 845-758-0400
Fax: 845-758-2633
Email: curran@proceedings.com
Web: www.proceedings.com

Table of contents

Volume 1035

Nonequilibrium thermodynamics and statistical physics: From rational modeling to its applications

16–17 March 2017, JR Hakata City, Fukuoka, Japan

Accepted papers received: 21 May 2018
Published online: 30 May 2018

Preface

Papers

IOP Conf. Series: Journal of Physics: Conf. Series **1035** (2018) 011001 doi:10.1088/1742-6596/1035/1/011001

Preface

The present volume contains 10 contributed papers presented at the meeting on *"Nonequilibrium thermodynamics and statistical physics: From rational modeling to its applications"* held during March 16 – 17, 2017 in Hakata, Japan. Each submitted contributed paper was peer-reviewed by at least one referee.

The meeting was dedicated to 66th birthday of Prof. Masaru Sugiyama in Nagoya Institute of Technology. The aim of this meeting is to provide the related researchers in the fields of nonequilibrium thermodynamics and statistical physics with an opportunity to get together and to exchange their recent results. Main topics of the meeting are the following:

- Rational extended thermodynamics
- Nonequilibrium statistical physics
- Kinetic theory of rarefied gases
- Nonlinear wave propagation
- Computational fluid dynamics
- Numerical simulation on material science

Valuable discussions from the viewpoints of theoretical physics, chemical physics and applied mathematics were also made after the talks.

This international meeting attracted 15 participants from 3 countries, namely, Japan, Italy and China. 12 oral contributions were presented at the meeting.

We would like to thank all participants for their essential contributions to the success of the meeting. We are also grateful to the reviewers for their careful reading and prompt response. The meeting was partially supported by JSPS KAKENHI Grant Numbers JP16K17555 (S. T.) JP15K21452 (T.A.) and by National Institute of Technology, Kitakyushu College, Japan.

Editors: Shigeru Taniguchi and Takashi Arima

IOP Conf. Series: Journal of Physics: Conf. Series **1035** (2018) 011002 doi:10.1088/1742-6596/1035/1/011002

Peer review statement

All papers published in this volume of *Journal of Physics: Conference Series* have been peer reviewed through processes administered by the proceedings Editors. Reviews were conducted by expert referees to the professional and scientific standards expected of a proceedings journal published by IOP Publishing.

IOP Conf. Series: Journal of Physics: Conf. Series **1035** (2018) 012001 doi:10.1088/1742-6596/1035/1/012001

On the Lindblad equation for open quantum systems: Rényi entropy rate and weak invariants

Sumiyoshi Abe[1,2,3]

[1] Physics Division, College of Information Science and Engineering, Huaqiao University, Xiamen 361021, China

[2] Department of Physical Engineering, Mie University, Mie 514-8507, Japan

[3] Institute of Physics, Kazan Federal University, Kazan 420008, Russia

E-mail: suabe@sf6.so-net.ne.jp

Abstract. A brief review of two recent topics on the Lindblad equation is presented. One is concerned with time evolution of the quantum Rényi entropy under the equation. The lower bound of the entropy rate is derived in a compact form. The other is about the concept of weak invariants, which generalize the Lewis-Riesenfeld invariant and are defined in such a way that they are not constant but their expectation values remain invariant in time. The Lindbladian operator describing the time-dependent damped quantum oscillator is identified, and the corresponding weak invariant is explicitly constructed as an example.

1. Introduction

There is growing *new interest* in quantum theory of open systems. This stream has originated from areas of contemporary importance such as nanoscience and quantum information. There, deeper understanding of quantum entanglement, decoherence and thermalization is desired.

Here, we discuss a couple of recent topics in quantum theory of open systems: time evolution of the quantum Rényi entropy and the concept of weak invariants. We develop discussions about them based on the Lindblad equation [1,2], which is the most general linear Markovian quantum master equation that preserves positive semidefiniteness of a density operator and generates a dynamical semigroup.

The present article is organized as follows. In Section 2, an elementary explanation is presented about the Lindblad equation. In Section 3, the general lower bound of the Rényi entropy rate under the Lindblad equation is discussed. In Section 4, the concept of weak invariants is formulated. As an example, the weak invariant of the time-dependent damped quantum oscillator is explicitly constructed. Section 5 is devoted to concluding remarks.

Throughout this article, $\hbar$ is set equal to unity for the sake of simplicity.

2. Lindblad equation

Consider an objective system A and the environment B that jointly form an isolated system. Assume that at certain time t the total system is in an *unentangled* pure state, $|\Psi(t)\rangle_{AB} = |\psi(t)\rangle_A |\phi(t)\rangle_B$. Time evolution of this state may be given by a transformation: $|\Psi(t)\rangle_{AB} \to |\Psi(t')\rangle_{AB} = U_{AB}(t',t)\,|\Psi(t)\rangle_{AB}$, where $U_{AB}(t',t)$ is a unitary time-evolution operator. U_{AB} is termed a local unitary operator if it has the form, $U_{AB} = U_A \otimes U_B$, where $U_A(U_B)$ is a unitary

IOP Conf. Series: Journal of Physics: Conf. Series **1035** (2018) 012001 doi:10.1088/1742-6596/1035/1/012001

operator in the space of the system A (B). Clearly, a local unitary time-evolution operator does not induce entanglement between A and B, and accordingly the marginal density operator of A, $\rho_A(t') = \mathrm{tr}_B[|\Psi(t')\rangle_{AB}\,{}_{AB}\langle\Psi(t')|]$, remains pure, where the symbol tr_B denotes the partial trace over B. Therefore, what is of interest is the case when U_{AB} is nonlocal. Nonlocality can be realized if there exists an interaction between A and B. Then, the total Hamiltonian, which generates time evolution of $|\Psi(t)\rangle_{AB}$, reads $H_{tot} = H_A + H_B + H_{AB}$, where $H_A(H_B)$ and H_{AB} are the Hamiltonians of A (B) and the interaction, respectively. The marginal density operator of A is written as follows:

$$\rho_A(t') = \sum_i V_i(t',t)\rho_A(t)V_i^\dagger(t',t), \tag{1}$$

where $\rho_A(t) = |\psi(t)\rangle_A\,{}_A\langle\psi(t)|$ is the pure-state density operator of A at t and V_i is the operator given by

$$V_i(t',t) = {}_B\langle v_i|U_{AB}(t',t)|\phi(t)\rangle_B \tag{2}$$

with $\{|v_i\rangle_B\}_i$ being a certain complete orthonormal system in the space of B. Since U_{AB} is unitary, $\mathrm{tr}_A\rho_A(t') = \mathrm{tr}_A\rho_A(t) = 1$, which leads to the trace-preserving condition

$$\sum_i V_i^\dagger V_i = I_A, \tag{3}$$

where I_A is the identity operator in the space of A.

Equation (1) is referred to as the Kraus representation [3] in the literature and gives an example of completely positive maps. The set $\{V_i^\dagger V_i\}_i$ satisfying Eq. (3) is called a positive operator-valued measure commonly abbreviated as POVM.

Let us consider infinitesimal time evolution from t to $t' = t + \Delta t$ [4]. We pick up one from V_i's, say V_0, and write it as

$$V_0(t+\Delta t,t) = I_A - i\Delta t H_A - \frac{\Delta t}{2}\sum_{i\neq 0}|g_i|^2 L_i^\dagger L_i + O((\Delta t)^2), \tag{4}$$

and the rest as

$$V_i(t+\Delta t,t) = (\Delta t)^{\frac{1}{2}} g_i L_i + O(\Delta t) \qquad (i\neq 0), \tag{5}$$

where g_i's are constant and the operators L_i's may depend on t. From these, we ascertain that Eq. (3) holds up to the first order of Δt.

Substituting Eqs. (4) and (5) into Eq. (1) with $t' = t + \Delta t$, we have

$$\rho_A(t+\Delta t) = \rho_A(t) - i\Delta t[H_A,\rho_A(t)] - \frac{\Delta t}{2}\sum_{i\neq 0}|g_i|^2\left(L_i^\dagger L_i\rho_A(t) + \rho_A(t)L_i^\dagger L_i - 2L_i\rho_A(t)L_i^\dagger\right) + O((\Delta t)^{3/2}), \tag{6}$$

from which we obtain in the limit $\Delta t \to 0$ the Lindblad equation [1,2]:

$$i\frac{\partial\rho}{\partial t} = [H,\rho] - \frac{i}{2}\sum_k c_k(L_k^\dagger L_k\rho + \rho L_k^\dagger L_k - 2L_k\rho L_k^\dagger), \tag{7}$$

which is local in time, i.e., Markovian. k runs over the set of i except $i = 0$ and $c_k = |g_k|^2$. Here and hereafter, the subscript "A" is omitted.

Nonnegativity of c_k's is crucial for the density operator to remain positive semidefinite. To see this point, let us perform instantaneous diagonalization:

$$\rho(t) = \sum_\lambda p_\lambda(t)|u_\lambda(t)\rangle\langle u_\lambda(t)|. \tag{8}$$

Here, $0 \leq p_\lambda(t) \leq 1$ and $\sum_\lambda p_\lambda(t) = 1$. $\{|u_\lambda(t)\rangle\}_\lambda$ is a complete orthonormal system, which consists of the eigenstates of $\rho(t)$ itself with the eigenvalues $p_\lambda(t)$'s. Suppose $p_1(t) = 0$ [5]. Then, the (1,1)-element of $\partial\rho(t)/\partial t$ is given by $\langle u_1(t)|\partial\rho(t)/\partial t|u_1(t)\rangle = dp_1(t)/dt$, provided that the condition $\langle u_\lambda(t)|u_{\lambda'}(t)\rangle = \delta_{\lambda\lambda'}$ has been used. Therefore, from Eq. (7), we have $dp_1/dt = \sum_k c_k \sum_{\lambda\neq 1} p_\lambda(t)|\langle u_1(t)|L_k|u_\lambda(t)\rangle|^2 \geq 0$, which guarantees that $p_1(t)$ never becomes negative.

3. Rényi entropy rate under Lindblad equation

Characterizing entanglement is of crucial importance in diverse areas in contemporary physics from quantum theory of spacetime [6] to quantum information [7]. In recent years, the quantum Rényi entropy has repeatedly been examined as a useful measure of entanglement. Considering the Rényi entropy rate may shed light on entanglement dynamics.

The quantum Rényi entropy of a density operator ρ is defined by

$$S_\alpha = \frac{1}{1-\alpha}\ln(\mathrm{tr}\,\rho^\alpha), \tag{9}$$

where α is a positive index. This quantity converges to the von Neumann entropy $S = -\,\mathrm{tr}(\rho\ln\rho))$ in the limit $\alpha \to 1$. α as a parameter plays a convenient role in probing entanglement (see [8-10], for example). In the diagonal form of the density operator in Eq. (8), Eq. (9) is rewritten as follows:

$$S_\alpha = \frac{1}{1-\alpha}\ln\left(\sum_\lambda p_\lambda^\alpha\right), \tag{10}$$

which is formally equivalent to the original expression of the classical Rényi entropy [11].

A couple of comments are made on the Rényi entropy. Firstly, not only the Rényi entropy but also any other generalized entropy cannot be the thermodynamic entropy. Maximum entropy methods with generalized entropies violate the Shore-Johnson axioms [12-14], and therefore artificial biases are introduced into such inference schemes [15,16]. Secondly, the Rényi entropy is as a functional not uniformly continuous in the limit of a large number of microscopically accessible states. This issue is concerned with the so-called Lesche stability [17] (see also [18]). It implies that the values of the Rényi entropy for two distributions slightly different from each other can largely be different.

Let us proceed to a discussion about time evolution of the Rényi entropy under the Lindblad equation. Taking the time derivative of the Rényi entropy and using Eq. (7), we have

$$\frac{dS_\alpha}{dt} = \frac{1}{\mathrm{tr}\,\rho^\alpha}\sum_k c_k G_k, \tag{11}$$

where

$$G_k = \frac{\alpha}{1-\alpha}\,\mathrm{tr}(\rho^{\alpha-1}L_k\rho L_k^\dagger - \rho^\alpha L_k^\dagger L_k). \tag{12}$$

Substituting the instantaneously diagonalized form of the density operator in Eq. (8) into Eq. (12), we have

$$G_k = \frac{\alpha}{1-\alpha}\left(\sum_{\lambda,\mu} p_\lambda^{\alpha-1}p_\mu|\langle u_\lambda|L_k|u_\mu\rangle|^2 - \sum_\lambda p_\lambda^\alpha\langle u_\lambda|L_k^\dagger L_k|u_\lambda\rangle\right). \tag{13}$$

Here and hereafter, the density operator is assumed to be positive definite. To further evaluate this quantity, we consider a function $f(x) = x^\alpha$ defined for positive x, which is concave (convex) if $0 < \alpha < 1$ ($\alpha > 1$). That is,

$$[rp_\lambda + (1-r)p_\mu]^\alpha > rp_\lambda^\alpha + (1-r)p_\mu^\alpha \qquad (0 < \alpha < 1), \tag{14}$$

IOP Conf. Series: Journal of Physics: Conf. Series **1035** (2018) 012001 doi:10.1088/1742-6596/1035/1/012001

$$[rp_\lambda + (1-r)p_\mu]^\alpha < rp_\lambda^\alpha + (1-r)p_\mu^\alpha \qquad (\alpha > 1), \tag{15}$$

where $0 < r < 1$. In addition, the following inequalities hold for $y > 0$: $y^\alpha - 1 \le \alpha(y-1)$ $(0 < \alpha < 1)$, $y^\alpha - 1 \ge \alpha(y-1)$ $(\alpha > 1)$, with the equalities for $y = 1$. Combining these with Eqs. (14) and (15), we have the following relations:

$$\alpha p_\lambda^{\alpha-1} p_\mu > (\alpha - 1)p_\lambda^\alpha + p_\mu^\alpha \qquad (0 < \alpha < 1), \tag{16}$$

$$\alpha p_\lambda^{\alpha-1} p_\mu < (\alpha - 1)p_\lambda^\alpha + p_\mu^\alpha \qquad (\alpha > 1), \tag{17}$$

from which we find that the quantity in Eq. (13) satisfies

$$G_k > \sum_\lambda p_\lambda^\alpha \langle u_\lambda | [L_k^\dagger, L_k] | u_\lambda \rangle \qquad (\alpha > 0). \tag{18}$$

Therefore, we obtain the following result [19]:

$$\frac{dS_\alpha}{dt} > \sum_k c_k \langle [L_k^\dagger, L_k] \rangle_\alpha . \tag{19}$$

The symbol appearing on the right-hand side stands for the escort average defined by

$$\langle A \rangle_\alpha = \frac{\operatorname{tr}(A\rho^\alpha)}{\operatorname{tr}\rho^\alpha}. \tag{20}$$

Thus, a positive rate is realized if the operators L_k's are normal, i.e., $[L_k^\dagger, L_k] = 0$.

The result in Eq. (19) generalizes the one for the von Neumann entropy presented in Ref. [20], which is reproduced in the limit $\alpha \to 1$.

Note that the escort average is not a quantum-mechanical average unless $\alpha \to 1$. As shown in Refs. [21-24], like the Rényi entropy, the escort average too fails to be Lesche stable.

4. Weak invariants of quantum dissipative systems

Here, we consider a situation that the Hamiltonian of the open objective system has an explicit time dependence: $H = H(t)$. Let $I(t)$ be a time-dependent observable. It is referred to as a weak invariant [25] if its average $\langle I(t) \rangle = \operatorname{tr}(I(t)\rho(t))$ is constant in time. Suppose $\rho(t)$ to be a solution of the Lindblad equation in Eq. (7). Then, $I(t)$ is a weak invariant, if it satisfies the following equation:

$$i\frac{\partial I(t)}{\partial t} - [H(t), I(t)] - \frac{i}{2}\sum_k c_k (L_k^\dagger L_k I(t) + I(t) L_k^\dagger L_k - 2L_k^\dagger I(t) L_k) = 0. \tag{21}$$

This generalizes the Lewis-Riesenfeld invariant $I_0(t)$ [26] (see also Ref. [27]) that obeys the equation:

$$i\frac{\partial I_0(t)}{\partial t} - [H(t), I_0(t)] = 0. \tag{22}$$

Henceforth, $I_0(t)$ satisfying Eq. (22) is referred to as a strong invariant.

A crucial difference between strong and weak invariants is that the eigenvalues of a strong invariant are constant in time, whereas those of a weak invariant are not.

A weak invariant is of physical relevance. For example, it is known [28-30] that there exists a process peculiar in quantum thermodynamics called an isoenergetic process, along which the internal energy is kept constant. Such a process is different from an isothermal process because of the quantum-mechanical violation of the law of equipartition of energy. If an isoenergetic process

IOP Conf. Series: Journal of Physics: Conf. Series **1035** (2018) 012001 doi:10.1088/1742-6596/1035/1/012001

is examined in the framework of finite-time quantum thermodynamics, then the time-dependent Hamiltonian is a weak invariant.

Strong invariants of the Lewis-Riesenfeld type have widely been applied to diverse problems in the literature. Examples include constructions of the coherent and squeezed states [31-34], geometric phases [35-37], nonstationary quantum field theory [38,39], fermions [40], quantum computation [41] and third-quantized cosmology [42].

The Lewis-Riesenfeld invariant is best known for the harmonic oscillator with the time-dependent frequency, the Hamiltonian of which reads

$$H(t) = \frac{1}{2}p^2 + \frac{1}{2}\omega^2(t)x^2, \tag{23}$$

provided that the mass has been set equal to unity for the sake of simplicity. The Lewis-Riesenfeld invariant associated with this Hamiltonian is given by

$$I_0(t) = \frac{1}{2}\left[(\sigma_0 p - \dot{\sigma}_0 x)^2 + \left(\frac{x}{\sigma_0}\right)^2\right], \tag{24}$$

where $\sigma_0 = \sigma_0(t)$ is a c-number real quantity satisfying the following auxiliary equation:

$$\ddot{\sigma}_0 + \omega^2(t)\sigma_0 = \frac{1}{\sigma_0^3}, \tag{25}$$

where the overdots denote time derivatives: $\dot{\sigma}_0 = d\sigma_0/dt$, $\ddot{\sigma}_0 = d^2\sigma_0/dt^2$. It is then straightforward to ascertain that $I_0(t)$ in Eq. (24) satisfies Eq. (22).

To generalize the strong invariant $I_0(t)$ to the weak invariant of the time-dependent damped harmonic oscillator, we consider the following single Lindbladian operator:

$$L \equiv L_1 = K_1 + a(t)K_2 + b(t)K_3. \tag{26}$$

Here, K_i's are given by $K_1 = p^2/2$, $K_2 = x^2/2$ and $K_3 = (px + xp)/2$ that satisfy the commutation relations: $[K_1, K_2] = -iK_3$, $[K_2, K_3] = 2iK_2$, $[K_3, K_1] = 2iK_1$, which is isomorphic to the $su(1,1)$ Lie algebra. $a(t)$ and $b(t)$ are c-number real quantities to be discussed below. A possible factor of K_1 is absorbed into $c \equiv c_1(t)$ in Eq. (7). Since L in Eq. (26) is Hermitian, Eqs. (7) and (21) become

$$i\frac{\partial \rho}{\partial t} = [H(t), \rho] - \frac{i}{2}c(t)[L, [L, \rho]], \tag{27}$$

$$i\frac{\partial I(t)}{\partial t} - [H(t), I(t)] - \frac{i}{2}c(t)[L, [L, I(t)]] = 0, \tag{28}$$

respectively, where $H(t)$ is the Hamiltonian given in Eq. (23).

The idea proposed in Ref. [25] is to require the classical equation of motion of the damped harmonic oscillator to hold as the expectation value:

$$\frac{d^2\langle x\rangle}{dt^2} + 2\kappa(t)\frac{d\langle x\rangle}{dt} + \Omega^2(t)\langle x\rangle = 0. \tag{29}$$

Clearly, there exists a nontrivial point behind this requirement. In classical theory, the friction term is indivisibly connected with an external noise term, usually in conformity with the fluctuation-dissipation relation. In the case of the linear system, the expectation value satisfies the equation that is similar to Eq. (29). A possible interpretation of Eq. (29) is that what corresponds to the noise may be the combination of the quantum fluctuations and the environmental effects.

Now, from Eq. (27) with Eqs. (23) and (26), the friction coefficient and the (modulated) frequency are found to be expressed as follows:

$$\kappa(t) = \frac{1}{2}c(t)\left[a(t) - b^2(t)\right], \tag{30}$$

$$\Omega^2(t) = \omega^2(t) + \kappa^2(t) + \dot{\kappa}(t). \tag{31}$$

The nonnegativity condition may be imposed on these quantities. The weak invariant satisfying Eq. (28) is found to have the form

$$I(t) = \frac{1}{2}\left[(\sigma p - \dot{\sigma} x)^2 + \left(\frac{x}{\sigma}\right)^2\right]. \tag{32}$$

This is formally identical to Eq. (24). However, the c-number quantity $\sigma = \sigma(t)$ here satisfies

$$\ddot{\sigma} - \kappa(t)\dot{\sigma} + \omega^2(t)\sigma = \frac{1}{\sigma^3}, \tag{33}$$

which is in contrast to Eq. (25). $c(t)$, $a(t)$ and $b(t)$ are related to $\sigma(t)$ as follows:

$$c(t) = \frac{8\kappa\sigma^4}{(\sigma\dot{\sigma})^2 + 4}, \tag{34}$$

$$a(t) = \frac{1}{2}\left(\frac{\dot{\sigma}}{\sigma}\right)^2 + \frac{1}{\sigma^4}, \tag{35}$$

$$b(t) = -\frac{\dot{\sigma}}{2\sigma}. \tag{36}$$

It should be noted that the friction terms in Eqs. (29) and (33) have the opposite signs. Thus, existence of the weak invariant is linked with such a time reversal structure.

5. Concluding remarks

In this article, we have made a short review of a couple of recent topics in quantum theory of Markovian open systems, i.e., time evolution of the Rényi entropy and weak invariants.

While this article has been under preparation, our attention has been drawn to Ref. [43]. There, the authors examine time evolution of the Rényi entropy for characterizing (non-) Markovianity of quantum subdynamics.

Clearly, weak invariants can also be defined for classical systems. For example, an weak invariant associated with the Fokker-Planck equation and its application to an econophysical problem have recently been discussed in Ref. [44]. A further investigation along this line is in progress.

Acknowledgments

This work has been supported in part by grants from National Natural Science Foundation of China (No. 11775084) and Grant-in-Aid for Scientific Research from the Japan Society for the Promotion of Science (No. 26400391 and No. 16K05484), and by the Program of Competitive Growth of Kazan Federal University from the Ministry of Education and Science of the Russian Federation.

Note Added. After completing the draft of the present paper, the author has noticed the work in Ref. [45], in which a weak invariant is discussed in the context of decoherence of the n-qubit system.

IOP Conf. Series: Journal of Physics: Conf. Series **1035** (2018) 012001 doi:10.1088/1742-6596/1035/1/012001

References

[1] Lindblad G 1976 *Commun. Math. Phys.* **48** 119
[2] Gorini V, Kossakowski A and Sudarshan E C G 1976 *J. Math. Phys.* **17** 821
[3] Kraus K 1983 States, *Effects, and Operations: Fundamental Notions of Quantum Theory* (Heidelberg: Springer-Verlag)
[4] Preskill J 2004 *Lecture Notes for Ph219/CS219: Quantum Information, Chapter 3* available at http://www.theory.caltech.edu/people/preskill/ph219/chap3_15.pdf
[5] Banks T, Susskind L and Peskin M E 1984 *Nucl. Phys.* B **244** 125
[6] Cowen R 2015 *Nature* **527** 290
[7] Nielsen M A and Chuang I L 2000 *Quantum Computation and Quantum Information* (Cambridge: Cambridge University Press)
[8] Jin B-Q and Korepin V E 2004 *J. Stat. Phys.* **116** 79
[9] Ares F, Esteve J G and Falceto F 2015 *Int. J. Geom. Methods Mod. Phys.* **12** 1560002
[10] Hamma A, Giampaolo S M and Illuminati F 2016 *Phys. Rev.* A **93** 012303
[11] Rényi A 2007 *Probability Theory* (New York: Dover)
[12] Shore J E and Johnson R W 1980 *IEEE Trans. Inf. Theory* **26** 26
[13] Shore J E and Johnson R W 1981 *IEEE Trans. Inf. Theory* **27** 472
[14] Johnson R W and Shore J E 1983 *IEEE Trans. Inf. Theory* **29** 942
[15] Pressé S, Ghosh K, Lee J and Dill K A 2013 *Phys. Rev. Lett.* **111** 180604
[16] Pressé S, Ghosh K, Lee J and Dill K A 2015 *Entropy* **17** 5043
[17] Lesche B 1982 *J. Stat. Phys.* **27** 419
[18] Abe S 2002 *Phys. Rev.* E **66** 046134
[19] Abe S 2016 *Phys. Rev.* E **94** 022106
[20] Benatti F and Narnhofer H 1988 *Lett. Math. Phys.* **5** 325
[21] Abe S 2008 *EPL* **84** 60006
[22] Abe S 2009 *J. Stat. Mech.: Theor. Exp.* **2009** P07027
[23] El Kaabouchi Aziz and Abe S 2011 *J. Appl. Math.* **2011** 804836
[24] Abe S 2012 *J. Phys.: Conf. Ser.* **394** 012003
[25] Abe S 2016 *Phys. Rev.* A **94** 032116
[26] Lewis Jr. H R and Riesenfeld W B 1969 *J. Math. Phys.* **10** 1458
[27] Dodonov V V and Man'ko V I 1978 *Physica* A **94** 403
[28] Ou C and Abe S 2016 *EPL* **113** 40009
[29] Liu S and Ou C 2016 *Entropy* **18** 205
[30] Ou C, Chamberlin R V and Abe S 2017 *Physica* A **466** 450
[31] Hartley J G and Ray J R 1982 *Phys. Rev.* D **25** 382
[32] Rajagopal A K and Marshall J T 1982 *Phys. Rev.* A **26** 2977
[33] Ma X 1988 *Phys. Rev.* A **38** 3548
[34] Abe S and Ehrhardt R 1993 *Phys. Rev.* A **48** 986
[35] Datta N, Ghosh G and Engineer M H 1989 *Phys. Rev.* A **40** 526
[36] Dittrich W and Reuter M 1991 *Phys. Lett.* A **155** 94
[37] Gao X-C, Xu J-B and Qian T-Z 1991 *Phys. Rev.* A **44** 7016
[38] Abe S and Ehrhardt R 1993 *Z. Phys.* C **57** 471
[39] Bertoni C, Finelli F and Venturi G 1998 *Phys. Lett.* A **237** 331
[40] Abe S 1993 *Phys. Lett.* A **181** 359
[41] Sarandy M S, Duzzioni E I and Serra R M 2011 *Phys. Lett.* A **375** 3343
[42] Abe S 1993 *Phys. Rev.* D **47** 718
[43] Song H and Mao Y 2017 *Phys. Rev.* A **96** 032115
[44] Abe S 2017 *Eur. Phys. J. Special Topics* **226** 529
[45] Wu S L, Zhang X Y and Yi X X 2015 *Phys. Rev.* A **92** 062122

IOP Conf. Series: Journal of Physics: Conf. Series **1035** (2018) 012002 doi:10.1088/1742-6596/1035/1/012002

Six-field extended thermodynamics models representing molecular energy exchange in a dense polyatomic gas

Takashi Arima

Department of Mechanical Engineering, Faculty of Engineering, Kanagawa University, Yokohama 221-8686, Japan

E-mail: arima@kanagawa-u.ac.jp

Abstract. In the framework of extended thermodynamics, we present three models of a dense polyatomic gases with 6 independent fields. The three models, respectively, represent the relaxation process due to the different molecular energy exchange among the translational, rotational and vibrational energies of a molecule. The characteristic velocities, the local exceptionality and the Maxwellian iteration are also studied. As an example, we study the six-field models of a van der Waals fluid.

1. Introduction

Nonequilibrium phenomena in polyatomic gases, such as absorption and dispersion of ultrasonic waves [1, 2] and shock waves [3], are characterized by the energy exchanges among the translational, rotational, and vibrational modes of a molecule [4]. The energy exchanges are related to the fundamental process for the dynamic (nonequilibrium) pressure [5, 6]. In the past, such phenomena are studied basing on the thermodynamic theory of relaxation processes of internal variables [7, 8, 9] in the framework of the thermodynamics of irreversible processes (TIP) [10] which relies essentially on the assumption of local thermodynamic equilibrium.

Recently, beyond the assumption of the local thermodynamic equilibrium, rational extended thermodynamics (hereafter referred to as ET) [11, 12, 13] has been developed. ET is a thermodynamic theory that adopts dissipative fluxes as independent fields in addition to the usual hydrodynamic fields motivated by the moment system of the kinetic theory [14]. The time evolution of the fields is governed by the system of balance equations which satisfy the Galilean invariance. The constitutive relations in local form are derived by the entropy principle and the causality principle.

ET was firstly proposed for *rarefied monatomic* gases [12]. Recently, the theory for *rarefied polyatomic* gases [13] has been established phenomenologically [15, 16, 17, 18] and kinetic-theoretically using the maximum entropy principle [19, 20, 21]. One of the noticeable features of these theories is that the almost all phenomenological coefficients are determined by adopting the thermal and caloric equations of state. The ET theory, in particular, the theory with 14 fields (ET_{14}) which includes the Navier-Stokes-Fourier theory as a special case has been successfully utilized to the study of linear waves [22, 23], shock waves [24, 25], heat conduction [26, 27], and light scattering [28].

In rarefied gases, the dynamic pressure can be identified with the energy exchange between the modes of molecular motion [18, 29]. In the case that the molecular internal modes are treated as a unit, the role of the dynamic pressure has been studied most clearly and simply by using the ET theory with 6 fields (ET_6): mass density, velocity, specific internal energy, and dynamic pressure [30, 31, 32]. This theory has one-to-one correspondence with the Meixner theory of a relaxation process [8, 9]. The nonlinear constitutive relations were obtained phenomenologically [33] and kinetic-theoretically [34, 35]. ET_6 is useful, for example, in the study of shock waves [32, 36], and the results of which have been confirmed by a kinetic-theoretical study [37, 38].

The ET_6 theory of *dense polyatomic* gases has been proposed by postulating the principle of duality between rarefied and dense gases [5]. The physical interpretation of this principle is summarized in [6]. In this theory, instead of the dynamic pressure, the nonequilibrium energies which play the role of the thermodynamic potential [29] are adopted as independent fields.

In spite of the successful establishment of ET theories of polyatomic gases, there remain nonequilibrium phenomena which are not described well by these theories. In such phenomena, all internal modes can not be treated as a unit. Only recently, for rarefied gases, the ET theory which describes the internal modes individually has been proposed [18]. In particular, the ET theory with two molecular relaxation processes for the rotational and vibrational modes (ET_7) has been constructed. As a limiting case of ET_7, three different ET_6 theories have been proposed depending on the relaxation process due to the energy exchange. The theory which describes the relaxation of the vibrational (rotational) energy individually is referred to ET_6^{KR} (ET_6^{KV}), and the previous ET_6 [30, 31, 33, 5, 39] in which the energies of the internal modes are treated as a unit is renamed ET_6^{RV}.

The purpose of the present paper is to propose the ET_6 theories of dense polyatomic gases, in particular, the dense gas theories of ET_6^{KR} and ET_6^{KV}, and to reveal their characteristic features. In the previous study of ET_6^{KR} and ET_6^{KV} [18], the field equations for the moments of the velocity distribution function were closed by means of the maximum entropy principle since the rarefied gas was considered. In the present paper, we adopt the phenomenological closure by means of the entropy principle with a model of the energy exchange. This is because no reliable kinetic theory exists for dense gases.

The paper is organized as follows: In Section 2, we present the system of balance equations of three ET_6 theories and the requirement of the entropy principle. With an energy-exchange model, the constitutive relations are determined. In Section 3, we study the characteristic velocities, local exceptionality and Maxwellian iteration of ET_6 theories. In Section 4, as an example, ET_6 theories of a nonpolytropic van der Waals fluid (vdW) are developed. The last section is devoted to the concluding remarks.

2. Closed system of field equations with 6 independent fields

2.1. Thermal and caloric equations of state

We study dense polyatomic gases with the Hamiltonian in the form:

$$H = H^{K+U} + H^R + H^V, \tag{1}$$

where H^{K+U} $(= K + U)$ is the Hamiltonian of the subsystem, say *$(K+U)$-system*, having the kinetic energy K of molecular translational motion and the intermolecular potential energy U, H^R is the one of the subsystem, say *R-system*, having the energy of the rotational motion of constituent molecules and H^V is the one of the subsystem, say *V-system*, having the energy of the vibrational motion of constituent molecules. For rarefied polyatomic gases, the intermolecular potential energy U is ignored.

The thermal and caloric equations of state in terms of the mass density ρ and the temperature

IOP Conf. Series: Journal of Physics: Conf. Series **1035** (2018) 012002 doi:10.1088/1742-6596/1035/1/012002

T are expressed as

$$p = \tilde{p}(\rho, T), \quad \varepsilon = \tilde{\varepsilon}(\rho, T), \tag{2}$$

where p is the pressure and ε is the specific internal energy. From (1), the pressure p can be divided into three parts: the pressure p^{K+U} due to H^{K+U}, the pressure p^R due to H^R and the pressure p^V due to H^V. Since $p^R = 0$ and $p^V = 0$, we have

$$p = p^{K+U}; \quad p^{K+U} = \tilde{p}^{K+U}(\rho, T). \tag{3}$$

Similarly we have

$$\varepsilon = \varepsilon^{K+U} + \varepsilon^R + \varepsilon^V; \quad \varepsilon^{K+U} = \tilde{\varepsilon}^{K+U}(\rho, T), \quad \varepsilon^R = \tilde{\varepsilon}^R(T), \quad \varepsilon^V = \tilde{\varepsilon}^V(T). \tag{4}$$

Note that ε^R and ε^V depend only on T [40]. We will use also the expression of the pressure in terms of ρ and ε^{K+U}:

$$p = \check{p}(\rho, \varepsilon^{K+U}) = \check{p}^{K+U}(\rho, \varepsilon^{K+U}), \tag{5}$$

where

$$p = \tilde{p}^{K+U}(\rho, T) = \check{p}^{K+U}(\rho, \tilde{\varepsilon}^{K+U}(\rho, T)). \tag{6}$$

2.2. System of balance equations of ET_6

A dissipative process in a polyatomic gas is mainly caused by the energy exchanges among the $(K + U)$, R and V subsystems [18, 5, 6]. Let $\mathcal{E}^{K+U}$, $\mathcal{E}^R$ and $\mathcal{E}^V$ denote the specific energies of $(K + U)$, R and V subsystems, respectively. In an equilibrium state, these energies coincide, respectively, with ε^{K+U}, ε^R and ε^V. The time evolutions of $\mathcal{E}^{K+U}$, $\mathcal{E}^R$ and $\mathcal{E}^V$ are studied by ET_7 which adopts ρ, v_i (velocity), $\mathcal{E}^{K+U}$, $\mathcal{E}^R$ and $\mathcal{E}^V$ as independent fields [18].

In [18], it has been revealed that the order of the relaxation times of the processes due to the energy exchanges is different and depends on the situations under study. In the present paper, let us study the relaxation process with a time scale in which an energy exchange between two of the three subsystems relaxes quickly and therefore such energy exchange can be ignored. Then these two subsystems can be regarded as a unit, say subsystem A, and the thermodynamic process is dominantly characterized by the time evolution of the remaining subsystem, say subsystem B, of which energy exchange with subsystem A is slow. Let $\mathcal{E}^A$ and $\mathcal{E}^B$ denote the energies of subsystems A and B, respectively. Then the specific internal energy ε can be divided as follows:

$$\varepsilon = \mathcal{E}^A + \mathcal{E}^B. \tag{7}$$

Note that we consider the processes that, as usual, the energy related to the molecular internal motion (rotational and/or vibrational modes) relaxes slowly. Therefore this kind of energy is involved in $\mathcal{E}^B$ and the energy of the molecular translational motion is involved in $\mathcal{E}^A$. As shown in Table 1, there are three possible choices of the subsystems A and B (see also [18]): (i) R and V systems are treated as a unit and relax slowly ((RV)-process), (ii) $(K+U)$ and R systems are treated as a unit and V-system relaxes slowly ((KR)-process), and (iii) $(K+U)$ and V systems treated as a unit and R-system relaxes slowly ((KV)-process). The energies $\mathcal{E}^A$ and $\mathcal{E}^B$ are expressed as the sum of the energies of constituent subsystems. For example, for (KR)-process, we have $\mathcal{E}^A = \mathcal{E}^{K+U+R} = \mathcal{E}^{K+U} + \mathcal{E}^R$.

Table 1. Three possible choices of the subsystem A and B

		Process	A	B
(i)	ET_6^{RV}	(RV)	$K+U$	$R+V$
(ii)	ET_6^{KR}	(KR)	$K+U+R$	V
(iii)	ET_6^{KV}	(KV)	$K+U+V$	R

ET_6^{RV}, ET_6^{KR} and ET_6^{KV} are the theories which describe the (RV), (KR) and (KV)-process, respectively. These ET_6 theories adopt ρ, v_i, $\mathcal{E}^A$ and $\mathcal{E}^B$ as independent fields and the system of field equations is expressed, as studied in [29], by

$$\begin{aligned}
&\frac{\partial \rho}{\partial t}+\frac{\partial}{\partial x_i}(\rho v_i)=0,\\
&\frac{\partial \rho v_j}{\partial t}+\frac{\partial}{\partial x_i}(\rho v_i v_j+\mathcal{P}\delta_{ij})=0,\\
&\frac{\partial}{\partial t}(2\rho\mathcal{E}^A+\rho v^2)+\frac{\partial}{\partial x_i}\left\{\left(2\rho\mathcal{E}^A+\rho v^2+2\mathcal{P}\right)v_i\right\}=\hat{P}_{ll},\\
&\frac{\partial}{\partial t}\left(2\rho\mathcal{E}^B\right)+\frac{\partial}{\partial x_i}\left(2\rho\mathcal{E}^B v_i\right)=-\hat{P}_{ll},
\end{aligned} \tag{8}$$

where $\mathcal{P}$ is the pressure, which is composed of the equilibrium pressure p and the dynamic pressure Π, and $\hat{P}_{ll}$ $(-\hat{P}_{ll})$ is the production of $\rho\mathcal{E}^A$ $(\rho\mathcal{E}^B)$. By the Galilean invariance of the system [41, 5, 29], the fluxes of $\rho\mathcal{E}^A$ and $\rho\mathcal{E}^B$ are determined strictly, and it is determined that the production term $\hat{P}_{ll}$ is a velocity-independent quantity. The first two equations are the conservation laws of mass and momentum. The third and forth equations are, respectively, the relaxation equation of the energies $\rho\mathcal{E}^A$ and $\rho\mathcal{E}^B$. Since sum of them represents the conservation law of the energy, the productions of $\rho\mathcal{E}^A$ and $\rho\mathcal{E}^B$ are chosen as in (8). For (RV)-process, the system (8) coincides with the previous dense gas theory [5].

The constitutive relation of $\mathcal{P}$ which is assumed to depend on the independent fields locally and instantaneously:

$$\mathcal{P}=\mathcal{P}(\rho,\mathcal{E}^A,\mathcal{E}^B) \tag{9}$$

is determined by the entropy principle with an energy-exchange model in Sec. 2.4. The explicit expression of $\hat{P}_{ll}$ is evaluated approximately by the Maxwellian iteration in Sec. 3.

Remark. In general, there is also the relaxation process due to the energy exchange between K and U. However, in the present analysis, we assume that K and U are treated as a unit. As discussed in [5], this assumption indicates that we consider the energy exchanges among H^{K+U}, H^R and H^V are the most dominant mechanism of the emergence of the dynamic pressure. In the monatomic-gas limit, this mechanism predicts that the dynamic pressure disappears.

2.3. Entropy principle

The entropy principle requires the following inequality [12, 13] with h $(=\rho\eta)$ being the entropy density (η: specific entropy), h_i the entropy flux ($h_i=hv_i+\varphi_i$: φ_i is the non-convective entropy flux), and Σ the entropy production:

$$\frac{\partial h}{\partial t}+\frac{\partial h_i}{\partial x_i}=\Sigma\geq 0 \tag{10}$$

IOP Conf. Series: Journal of Physics: Conf. Series **1035** (2018) 012002 doi:10.1088/1742-6596/1035/1/012002

for any solution of the system. We assume that h, φ_i and Σ are constitutive quantities:

$$\eta = \eta(\rho, \mathcal{E}^A, \mathcal{E}^B), \qquad \varphi_i = \varphi_i(\rho, \mathcal{E}^A, \mathcal{E}^B), \qquad \Sigma = \Sigma(\rho, \mathcal{E}^A, \mathcal{E}^B). \tag{11}$$

As the result of the exploitation of the entropy principle, we obtain the following partial differential equation for η [5, 29]:

$$\eta_\rho + \frac{\mathcal{P}}{\rho^2}\eta_{\mathcal{E}^A} = 0, \tag{12}$$

where a subscript attached to η indicates a partial derivative. The nonequilibrium temperatures ϑ and Θ which are, respectively, related to $\mathcal{E}^A$ and $\mathcal{E}^B$ (see (21)) can be introduced through the nonequilibrium entropy density as follows [5, 13, 18, 29]:

$$\frac{1}{\vartheta} = \eta_{\mathcal{E}^A}, \qquad \frac{1}{\Theta} = \eta_{\mathcal{E}^B}. \tag{13}$$

In equilibrium, ϑ and Θ coincide with the equilibrium temperature T since $\mathcal{E}^A$ and $\mathcal{E}^B$ coincide with ε^A and ε^B. By using (12) and (13), the derivative of the nonequilibrium entropy density is obtained as follows:

$$\mathrm{d}\eta = -\frac{\mathcal{P}}{\rho^2\vartheta}\mathrm{d}\rho + \frac{1}{\vartheta}\mathrm{d}\mathcal{E}^A + \frac{1}{\Theta}\mathrm{d}\mathcal{E}^B. \tag{14}$$

The entropy principle also requires that the entropy production satisfies the following relation [5, 29]:

$$\Sigma = \frac{1}{2}\left(\frac{1}{\vartheta} - \frac{1}{\Theta}\right)\hat{P}_{ll} \geq 0. \tag{15}$$

In addition, from the representation theorem, we have $\varphi_i = 0$.

2.4. A model of nonequilibrium entropy and pressure

Except for the production term, there remain two constitutive quantities, that is, η and $\mathcal{P}$ which are related to each other by the relation (12). In the previous studies of ET_6^{RV}, two equivalent models to determine the constitutive relations have been proposed. The one is a model of the pressure [5, 6] which assumes that the dynamic pressure is caused by the energy exchange between the subsystems. The other one is a model of the nonequilibrium entropy density [29] which assumes on the basis of the energy-exchange model that η is the sum of the entropy densities of the subsystems.

By generalizing the latter model, we propose the following nonequilibrium entropy density:

$$\eta = s^A(\rho, \mathcal{E}^A) + s^B(\mathcal{E}^B), \tag{16}$$

where s^A and s^B are the nonequilibrium entropy densities of the subsystem which satisfy the Gibbs relations of each subsystem:

$$\vartheta \mathrm{d}s^A = \mathrm{d}\mathcal{E}^A - \frac{\check{p}(\rho, \mathcal{E}^A)}{\rho^2}\mathrm{d}\rho, \quad \Theta \mathrm{d}s^B = \mathrm{d}\mathcal{E}^B \tag{17}$$

with the nonequilibrium temperatures defined in (13). Note that s^B depends only on $\mathcal{E}^B$ because the subsystem B is related to only the internal modes. The nonequilibrium temperatures are expressed in this case as follows:

$$\frac{1}{\vartheta} = s^A_{\mathcal{E}^A}, \qquad \frac{1}{\Theta} = s^B_{\mathcal{E}^B}. \tag{18}$$

From (12), we obtain

$$\mathcal{P} = \breve{p}(\rho, \mathcal{E}^A), \tag{19}$$

then we also obtain the dynamic pressure as follows:

$$\Pi = \mathcal{P} - p = \breve{p}(\rho, \mathcal{E}^A) - \breve{p}(\rho, \varepsilon^A). \tag{20}$$

Let us adopt $\{\rho, v_i, \vartheta, \Theta\}$ as independent variables instead of $\{\rho, v_i, \mathcal{E}^A, \mathcal{E}^B\}$. From (18), $\mathcal{E}^A$ depends on ρ and ϑ, while $\mathcal{E}^B$ depends only on Θ. Since $\mathcal{E}^A$ and $\mathcal{E}^B$, respectively, coincide with $\varepsilon^A = \tilde{\varepsilon}^A(\rho, T)$ and $\varepsilon^B = \tilde{\varepsilon}^B(T)$ in equilibrium in which ϑ and Θ coincide with T, we notice that $\mathcal{E}^A$ and $\mathcal{E}^B$ follow the caloric equations of state of each subsystem:

$$\mathcal{E}^A = \tilde{\varepsilon}^A(\rho, \vartheta), \quad \mathcal{E}^B = \tilde{\varepsilon}^B(\Theta). \tag{21}$$

From (7), the total internal energy ε is given by

$$\tilde{\varepsilon}(\rho, T) = \tilde{\varepsilon}^A(\rho, \vartheta) + \tilde{\varepsilon}^B(\Theta). \tag{22}$$

This gives the relationship among the three temperatures T, ϑ, and Θ. From (20), the dynamic pressure is expressed as follows:

$$\Pi = \tilde{p}(\rho, \vartheta) - \tilde{p}(\rho, T). \tag{23}$$

The entropy density is restricted by the thermodynamic stability condition, which requires that the entropy must be convex with respect to the densities. In the present case, this condition is given by [29]

$$\tilde{p}_\rho(\rho, \vartheta) > 0, \quad c_v^A(\rho, \vartheta) > 0, \quad c_v^B(\Theta) > 0, \tag{24}$$

where c_v^A and c_v^B are the specific heats of the two subsystems defined by $c_v^A(\rho, \vartheta) = \partial\tilde{\varepsilon}^A(\rho, \vartheta)/\partial\vartheta$ and $c_v^B(\Theta) = \mathrm{d}\tilde{\varepsilon}^B(\Theta)/\mathrm{d}\Theta$.

Remark. Instead of the model of the entropy density (16), it is possible to adopt a model of the pressure. In [18], for rarefied gases, the expression of $\mathcal{P}$ is obtained in the form of (19) without the contribution of U. The principle of duality between rarefied and dense gases [5] states that the nonequilibrium pressure is caused by the energy exchange between the subsystems whether the contribution of U exists or not. In other words, the pressure evaluated with K in rarefied gases is replaced by the pressure evaluated with $(K + U)$ in dense gases. From this principle, we have (19) in dense gases. Then, we obtain (16) as a result.

3. Characteristic velocities, local exceptionality and Maxwellian iteration

The characteristic velocities $\mathcal{V}$ associated to the hyperbolic system (8) are obtained as

$$\mathcal{V} = v_n \,(\text{multiplicity } 4),\; v_n \pm \mathcal{U} \quad \text{with} \quad \mathcal{U}^2 = p_\rho(\rho, \vartheta) + \frac{\vartheta p_\vartheta^2(\rho, \vartheta)}{\rho^2 c_v^A(\rho, \vartheta)}, \tag{25}$$

where v_n is the velocity in normal direction to the wave front. Note that $\mathcal{U}$ satisfies the subcharacteristic condition [42] with respect to the Euler system. This condition ensures that ET_6 includes the Euler system as a principal subsystem.

In the theory of the hyperbolic system, the local exceptionality (locally linearly degenerate) hyper-surface is crucial to study the admissibility of the shock wave, in particular, rarefaction shock [43]. It is known that the sound waves can be locally exceptional if $\delta\mathcal{V} = 0$ for some states. In the present case, when $\{\rho, v_i, \eta, \Theta\}$ are adopted as independent fields, the hyper-surface of local exceptionality exists if the following relation is satisfied [5, 18]:

$$\left(\frac{\partial \rho^2 \mathcal{U}^2}{\partial \rho}\right)_{\eta,\Theta} = 0. \tag{26}$$

The energy exchange between two subsystems is characterized by the following quantity:

$$\Delta \equiv \mathcal{E}^A - \varepsilon^A = -\mathcal{E}^B + \varepsilon^B.$$

The linearized system around an equilibrium state (ρ, T) is obtained by expanding the system with respect to the energy exchange Δ up to the first order. The nonequilibrium temperatures are in terms of Δ as follows:

$$\vartheta - T = \frac{\Delta}{c_v^A}, \qquad \Theta - T = -\frac{\Delta}{c_v^B}, \tag{27}$$

and, from (23), the dynamic pressure is expressed as follows:

$$\Pi = p_T(\vartheta - T) = \frac{p_T}{c_v^A}\Delta. \tag{28}$$

Here and hereafter we use the notation c_v^A instead of $c_v^A(\rho, T)$ and so on for simplicity. The production term is given by

$$\hat{P}_{ll} = -\frac{2\rho}{\tau}\Delta, \tag{29}$$

where $\tau = \tau(\rho, T)$ is the relaxation time for Δ which is positive from the requirement of the entropy principle (15). It is noted that τ is also interpreted as the relaxation time of the dynamic pressure which is studied in previous studies [15, 22, 24, 30, 32]. The system of field equations (8) is rewritten with the independent fields $\{\rho, v_i, T, \Delta\}$ as follows:

$$\begin{aligned}
&\dot{\rho} + \rho\frac{\partial v_i}{\partial x_i} = 0,\\
&\rho\dot{v}_i + \frac{\partial}{\partial x_i}\left(p + \frac{p_T}{c_v^A}\Delta\right) = 0,\\
&\rho c_v \dot{T} + p_T\left(T + \frac{\Delta}{c_v^A}\right)\frac{\partial v_i}{\partial x_i} = 0,\\
&\dot{\Delta} + \frac{p_T c_v^B}{\rho c_v}\left(T + \frac{\Delta}{c_v^A}\right)\frac{\partial v_i}{\partial x_i} = -\frac{1}{\tau}\Delta,
\end{aligned} \tag{30}$$

where c_v is the specific heat defined by $c_v(\rho, T) = \partial\tilde{\varepsilon}(\rho, T)/\partial T$ which satisfies $c_v = c_v^A + c_v^B$. By adopting the Maxwellian iteration [44], we obtain the bulk viscosity ν in terms of τ:

$$\nu = \tau\frac{T p_T^2}{\rho}\frac{c_v^B}{c_v c_v^A}. \tag{31}$$

The relaxation time τ can be estimated by using the experimental data on ν.

IOP Conf. Series: Journal of Physics: Conf. Series **1035** (2018) 012002 doi:10.1088/1742-6596/1035/1/012002

4. ET$_6$ theories of a van der Waals Fluid

4.1. Equations of state and dynamic pressure

Let us study, as a typical example, a nonpolytropic vdW fluid of which thermal and caloric equations of state are given by

$$p = \frac{k_B}{m}\frac{T\rho}{1-b\rho} - a\rho^2, \quad \varepsilon = \varepsilon_r - a\rho, \tag{32}$$

where ε_r is the specific internal energy of a rarefied gas and the material-dependent constants a and b represent, respectively, a measure of the strength of the attraction between constituent molecules and the effective volume (or exclusion volume) of a molecule. The specific internal energies of subsystems ε^A and ε^B are expressed with those in rarefied gases as follows:

$$\varepsilon^A = \tilde{\varepsilon}_r^A(T) - a\rho, \quad \varepsilon^B = \tilde{\varepsilon}_r^B(T). \tag{33}$$

In particular, since the specific heat of intermolecular potential is zero, we have

$$c_v^A(T) = c_{v_r}^A(T), \quad c_v^B(T) = c_{v_r}^B(T), \tag{34}$$

where $c_{v_r}^{A,B}$ are the specific heats of the subsystem in a rarefied gas.

From (23), the dynamic pressure is expressed as follows:

$$\Pi = \frac{k_B}{m}\frac{\rho}{1-b\rho}(\vartheta - T). \tag{35}$$

The relation between the bulk viscosity and the relaxation time (31) is explicitly obtained as follows:

$$\nu = \frac{k_B}{m}\frac{c_v^B}{c_v c_v^A}\frac{T\rho}{(1-b\rho)^2}\tau. \tag{36}$$

4.2. Nonequilibrium entropy and its convexity in a vdW fluid

From (16) with the equilibrium entropy density of a vdW fluid, the nonequilibrium entropy density η is obtained as follows:

$$\eta = \frac{k_B}{m}\log\left(\frac{1-b\rho}{\rho}\right) + \int^{\vartheta}\frac{c_v^A(T')}{T'}\mathrm{d}T' + \int^{\Theta}\frac{c_v^B(T')}{T'}\mathrm{d}T' + s_0, \tag{37}$$

where s_0 is a constant at a reference state.

As the conditions $(24)_{2,3}$ are identically satisfied, the convexity condition of the entropy density comes only from $(24)_1$ and is expressed as

$$\hat{p}(\hat{\rho},\hat{\vartheta}) > (3-2\hat{\rho})\hat{\rho}^2, \tag{38}$$

where we have introduced the following dimensionless variables:

$$\hat{p} = \frac{p}{p_{cr}}, \; \hat{\rho} = \frac{\rho}{\rho_{cr}}, \; \hat{\vartheta} = \frac{\vartheta}{T_{cr}}, \; \hat{\Theta} = \frac{\Theta}{T_{cr}}, \tag{39}$$

with $p_{cr} = a/(27b^2)$, $\rho_{cr} = 1/(3b)$ and $T_{cr} = 8a/(27\frac{k_B}{m}b)$ which are the quantities at the critical point. In the $(\hat{\rho}, \hat{p}(\hat{\rho},\hat{\vartheta}))$-plane, the boundary (spinodal curve) is independent of $\hat{\vartheta}$ as similar to the case of the Euler system in which the spinodal curve for equilibrium pressure is independent of the temperature.

The critical point is the point on the spinodal curve with the condition that the second derivative also vanishes:

$$\left(\frac{\partial^2 p(\rho,\vartheta)}{\partial\rho^2}\right)_\vartheta = 0.$$

With (38), we have, as the Euler system,

$$\hat\rho = 1, \quad \hat\vartheta = 1, \quad \hat p(\hat\rho,\hat\vartheta) = 1. \tag{40}$$

Remark. The expressions of the spinodal curve and critical point are independent from the molecular relaxation processes. However, ϑ depends on the process (see (27) near equilibrium case).

4.3. Characteristic velocity and critical derivative in a vdW fluid

In the present case, the characteristic velocity $\mathcal{U}$ is obtained as follows:

$$\mathcal{U}^2 = \frac{k_B}{m}\frac{\vartheta\gamma^A(\vartheta)}{(1-b\rho)^2} - 2a\rho, \tag{41}$$

where $\gamma^A(\vartheta) = \frac{\hat c_v^A(\vartheta)+1}{\hat c_v^A(\vartheta)}$ is the ratio of the specific heats of subsystem A with the dimensionless specific heat $\hat c_v^A = mc_v^A/k_B$.

The locus of the local exceptionality (26), called critical derivative curve (see [43]), is expressed as, in the dimensionless form with (39),

$$\hat p(\hat\rho,\hat\vartheta) = \hat\rho^2\left\{\frac{2(3-\hat\rho)^2}{\gamma^A(\vartheta)(\gamma^A(\vartheta)+1) - \frac{\vartheta}{\hat c_v^{A^3}(\vartheta)}\frac{\mathrm{d}\hat c_v^A(\vartheta)}{\mathrm{d}\vartheta}} - 3\right\}. \tag{42}$$

In comparison with the spinodal curve (38), the condition that the curve of local exceptionality resides in the stable region is

$$\frac{3-\hat\rho}{\gamma^A(\vartheta)(\gamma^A(\vartheta)+1) - \frac{\vartheta}{\hat c_v^{A^3}(\vartheta)}\frac{\mathrm{d}\hat c_v^A(\vartheta)}{\mathrm{d}\vartheta}} \geq 1. \tag{43}$$

For polytropic gases where the specific heats are constant, recalling $\gamma^A > 1$, the condition (43) can be expressed as

$$1 < \gamma^A \leq \sqrt{\frac{13}{4} - \hat\rho} - \frac{1}{2}. \tag{44}$$

The necessary condition that the locus resides in stable region is $\hat\rho < 1$. The condition (44) can be rewritten with respect to $\hat c_v^A$ as follows:

$$\hat c_v^A \geq \frac{1}{\sqrt{\frac{13}{4} - \hat\rho} - \frac{3}{2}} \geq \frac{3+\sqrt{13}}{2}. \tag{45}$$

As previously studied in [5], in the case of ET_6^{RV} in which $\hat c_v^A = 3/2$, it is observed that the locus (42) is independent of the nonequilibrium variables. In this case the locus of local exceptionality

IOP Conf. Series: Journal of Physics: Conf. Series **1035** (2018) 012002 doi:10.1088/1742-6596/1035/1/012002

coincides with the curve for a monatomic Euler fluid and always resides in the unstable region. On the other hand, in the cases of ET_6^{KR} and ET_6^{KV} in which $\hat{c}_v^A > 3/2$, with a large value of $\hat{c}_v^A$, there exists a region in the $(\hat{\rho}, \hat{p}(\hat{\rho}, \hat{\vartheta}))$-plane where the local exceptionality curve exists in the stable region. Therefore, in such cases, the rarefaction shock may appear. The same feature is observed in a polyatomic Euler system [43]. The loci of the local exceptionality in the cases that $\hat{c}_v^A = 3/2$, 7/2 and 11/2 are shown in Figure 1.

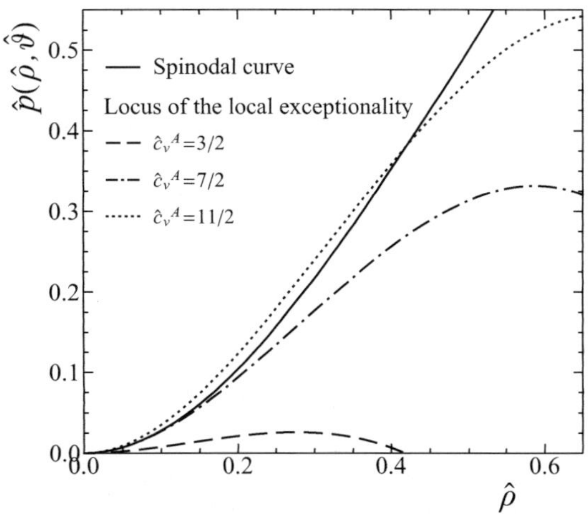

Figure 1. Loci of the local exceptionality (42) with $\hat{c}_v^A = 3/2$ (dashed line), 7/2 (dot-dashed line) and 11/2 (dotted line) and spinodal curve (solid line) in the $(\hat{\rho}, \hat{p}(\hat{\rho}, \hat{\vartheta}))$-plane.

5. Summary and concluding remarks

We have studied three ET_6 theories of dense polyatomic gases in which the internal modes of a molecule play a significant role. These theories are based on three different relaxation processes due to the energy exchange between the subsystems. The model of the nonequilibrium entropy density is proposed on the basis of the energy-exchange model and the system of the field equations is phenomenologically closed. The characteristic velocities, the local exceptionality and the Maxwellian iteration have been studied. As an example, ET_6 theories of a nonpolytropic vdW fluid is constructed. In particular, it has been elucidated that, in a polytropic vdW fluid, the locus of the local exceptionality of ET_6^{KR} and ET_6^{KV} resides in the stable region with a large value of specific heat of the subsystem A.

Finally, three concluding remarks are made:

(i) The present procedure to obtain the dense gas theory can be applied to the case of ET_7. Then the relaxation processes of rotational and vibrational modes are described separately. In ET_7 of a polytropic vdW fluid, the existence of the locus of the local exceptionality in the stable region may be suspicious because its characteristic velocities are same as those of ET_6^{RV}. These studies will be the subject of the next paper.

(ii) There remains a question how to determine the relaxation process under study. One way is to choose the process which is the most appropriate in comparison with the experimental data. The comparison of the theoretical prediction of the dispersion relation of a sound wave in dense gases by ET with the experimental data will soon be reported elsewhere.

(iii) From an experimental point of view, it is well known that the existence of the rarefaction shock is still controversial. Since the present analysis is based on the simplified theory, the study of the admissibility of the rarefaction shock by a more realistic ET theory with shear stress and heat flux is the subject for subsequent works.

Acknowledgments

The present paper is dedicated with deep esteem and gratitude to Prof. Masaru Sugiyama on the occasion of his 66th birthday. The results contained in the present paper have been partially presented in "Nonequilibrium thermodynamics and statistical physics: From rational modeling to its applications". This work was supported by JSPS KAKENHI Grant No. JP15K21452.

References

[1] Herzfeld K F and Litovitz T A 1959 *Absorption and dispersion of ultrasonic waves* (New York, London: Academic Press)
[2] Bhatia A B 1985 *Ultrasonic absorption: An introduction to the theory of sound absorption and dispersion in gases, liquids, and solids* (New York: Dover)
[3] Vincenti W G and Kruger C H 1965 *Introduction to Physical Gas Dynamics* (New York, London, Sydney: John Wiley and Sons)
[4] Herzfeld K F and Rice F O 1928 *Phys. Rev.* **31** 691
[5] Arima T, Ruggeri T and Sugiyama M 2017 *Phys. Rev. Fluids* **2** 013401
[6] Arima T and Sugiyama M *Ricerche mat.* (*Preprint* https://doi.org/10.1007/s11587-018-0386-8)
[7] Mandelstam L I and Leontovich M A 1937 *Zh. Eksp. Teor. Fiz.* **7** 438
[8] Meixner J 1943 *Ann. Phys.* **43** 470
[9] Meixner J 1952 *Acoustica* **2** 101
[10] de Groot S R and Mazur P 1984 *Non-Equilibrium Thermodynamics* (New York: Dover)
[11] Müller I and Ruggeri T 1993 *Extended Thermodynamics* 1st ed. (New York: Springer)
[12] Müller I and Ruggeri T 1998 *Rational Extended Thermodynamics* 2nd ed. (New York: Springer)
[13] Ruggeri T and Sugiyama M 2015 *Rational Extended Thermodynamics beyond the Monatomic Gas* (Cham, Heidelbergh, New York, Dorderecht, London: Springer)
[14] Grad H 1949 *Comm. Pure Appl. Math.* **2** 331
[15] Arima T, Taniguchi S, Ruggeri T and Sugiyama M 2011 *Continuum Mech. Thermodyn.* **24** 271
[16] Arima T and Sugiyama M 2013 *Atti della Accademia Peloritana dei Pericolanti* **91(S1)** A1
[17] Ruggeri T and Strumia A 1981 *Ann. Inst. H. Poincaré, Section A* **34** 65
[18] Arima T, Ruggeri T and Sugiyama M 2017 *Phys. Rev. E* **96** 042143
[19] Pavić M, Ruggeri T and Simić S 2013 *Physica A* **392** 1302
[20] Arima T, Mentrelli A and Ruggeri T 2014 *Ann. Phys.* **345** 111
[21] Arima T, Mentrelli A and Ruggeri T 2014 *Rend. Lincei Mat. Appl.* **25** 275
[22] Arima T, Taniguchi S, Ruggeri T and Sugiyama M 2013 *Continuum Mech. Thermodyn.* **25** 727
[23] Arima T, Taniguchi S, Ruggeri T and Sugiyama M 2014 *Acta Appl. Math.* **132** 15
[24] Taniguchi S, Arima T, Ruggeri T and Sugiyama M 2014 *Phys. Rev. E* **89** 013025
[25] Taniguchi S, Arima T, Ruggeri T and Sugiyama M 2014 *Acta Appl. Math.* **132** 583
[26] Arima T, Barbera E, Brini F and Sugiyama M 2014 *Phys. Lett. A* **378** 2695
[27] Barbera E, Brini F and Sugiyama M 2014 *Acta Appl. Math.* **132** 41
[28] Arima T, Taniguchi S and Sugiyama M 2013 *Proc. 34th Symp. Ultrasonic Electronics* 15
[29] Arima T and Sugiyama M *Ricerche mat.* (*Preprint* https://doi.org/10.1007/s11587-018-0399-3)
[30] Arima T, Taniguchi S, Ruggeri T and Sugiyama M 2012 *Phys. Lett. A* **376** 2799
[31] Arima T, Ruggeri T, Sugiyama M and Taniguchi S 2014 *Meccanica* **49** 2181
[32] Taniguchi S, Arima T, Ruggeri T and Sugiyama M 2014 *Phys. Fluids* **26** 016103
[33] Arima T, Ruggeri T, Sugiyama M and Taniguchi S 2015 *Int. J. Non-Linear Mech.* **72** 6
[34] Ruggeri T 2016 *Bull. Inst. Math. Acad. Sin.* **11** 1
[35] Bisi M, Ruggeri T and Spiga G 2018 *Kinet. Relat. Mod.* **11** 71
[36] Taniguchi S, Arima T, Ruggeri T and Sugiyama M 2016 *Int. J. Non Linear Mech.* **79** 66
[37] Kosuge S, Aoki K and Goto T 2016 *AIP Conf. Proc.* **1786** 180004
[38] Kosuge S and Aoki K 2018 *Phys. Rev. Fluids* **3** 023401
[39] Carrisi, M.C., Pennisi, S., Ruggeri, T., Sugiyama, M. 2015 *Ricerche mat.* **64** 403
[40] Landau L D and Lifshitz E M 1980 *Statistical Physics* (Oxford: Pergamon)
[41] Ruggeri T 1989 *Continuum Mech. Thermodyn.* **1** 3
[42] Boillat G and Ruggeri T 1997 *Arch. Rational Mech. Anal.* **137** 305
[43] Zhao N, Mentrelli A, Ruggeri T and Sugiyama M 2011 *Phys. Fluids* **23** 086101

[44] Ikenberry E and Truesdell C 1956 *J. Rational Mech. Anal.* **5** 1

IOP Conf. Series: Journal of Physics: Conf. Series **1035** (2018) 012003 doi:10.1088/1742-6596/1035/1/012003

Some 2D heat transfer problems in metals described by a linearized extended thermodynamics model

Elvira Barbera[1] **and Francesca Brini**[2]

[1] Department of Mathematical, Computer, Physical and Earth Sciences, University of Messina, Messina, Italy

[2] Department of Mathematics and Alma Mater Research Center on Applied Mathematics University of Bologna, Bologna, Italy

E-mail: `francesca.brini@unibo.it`

Abstract. We describe qualitatively some 2D stationary problems for metal electrons in the presence of a temperature gradient and a magnetic field. To this aim we refer to a linearized Extended Thermodynamics model that allows a semianalytical construction of the solutions. The results are in agreement with the thermomagnetic effects already known in the literature.

1. Introduction

In this paper we consider a metallic plate in the presence of a temperature gradient and of a magnetic flux density. In order to describe the stationary physical phenomena related to this case, we refer to a Rational Extended Thermodynamics (RET) model for free electrons bounded within a metal, in agreement with the original idea by Sommerfeld [1, 2, 3, 4, 5]. The electrons in the metallic body are described as free fermion particles of mass m. Electrons can collide occasionally with a lattice ion (mass M), but their energy is unchanged by a collision, due to the large ratio M/m. In fact, the ions are modeled as rigid spheres, at rest at their lattice point and their density is uniform and constant. Moreover, the collision between electrons are neglected.

Our analysis of the physical system will be developed within the the framework of RET [4, 6], a macroscopic theory based on a different strategy with respect to Classical Thermodynamics (CT). In fact, RET considers as field variables not only those of CT (mass density, momentum and energy) but also the stress tensor, the heat flux and others. The corresponding field equations are balance laws supplemented by local and instantaneous constitutive equations, that are determined by the requirement of validity of universal physical principles, like the entropy principle and the principle of relativity. During the last decades, RET proved to be a very powerful theory, capable of describing non-stationary physical phenomena through hyperbolic PDE systems, overcoming the paradox of infinite velocities due to parabolic PDE models. At the beginning, the theory was proposed by Müller, Ruggeri and other researchers for rarefied monatomic gases [4], but in the last years it has been generalized to rarefied polyatomic gases both in the classical [6, 7] and in the relativistic framework [8] and also to quantum systems [9], obtaining relevant results and very good agreement with experimental data (see for example [4, 6, 10, 11]). So, it is natural to turn to this theory also in the present case.

The physical phenoma that we will study here are basically associated to heat transfer and we will show the relevance of the boundary conditions assignment, in contrast to what happens for a classical gas. In particular, a special attention will be devoted to the Righi-Leduc effect.

The structure of the paper is the following. Section 2 is devoted to the equation model, while the physical problems is presented and analysed in Section 3. Some concluding remarks are proposed in Section 4.

2. The model

In this section we refer to the RET 8-moment model obtained by Müller [3, 4] through a microscopic approach and recently revisited in [5] at a macroscopic level. In order to derive this set of balance equations, following [1, 2], it is assumed that the state of the electron gas is described by the phase density $f(\mathbf{x}, \mathbf{c}, t)$ of electrons at the position $\mathbf{x}$, with the velocity $\mathbf{c}$ and at the time t. The phase density must satisfy the Boltzmann equation, which assumes the form

$$\frac{\partial f}{\partial t} + c_i \frac{\partial f}{\partial x_i} + f_i \frac{\partial f}{\partial c_i} = S, \tag{1}$$

where S stands for the collision term, while f_i represents the specific external force acting on the particles which, in a metal gas, is the Lorentz force

$$f_i = -\frac{q}{m}\left[E_i + (\mathbf{c} \wedge \mathbf{B})_i\right], \tag{2}$$

if $-q$ denotes the charge of an electron, m denotes its mass, $\mathbf{E}$ is the electromotive intensity and $\mathbf{B}$ represents the magnetic flux density, while '$\wedge$' is the symbol of the vector product.

Multiplying equations (1) by a generic function $\varphi(\mathbf{x}, \mathbf{c}, t)$ and integrating it over the whole range of $\mathbf{c}$, it is possible to construct the generic moment equation for the electron gas:

$$\frac{\partial \bar{\varphi}}{\partial t} + \frac{\partial \overline{c_i \varphi}}{\partial x_i} - \overline{\frac{\partial \varphi}{\partial t} + c_i \frac{\partial \varphi}{\partial x_i} + (f_i + i_i^c)\frac{\partial f}{\partial c_i}} = \int \varphi S \mathrm{d}\mathbf{c}. \tag{3}$$

where the symbol $\bar{\varphi}$ indicates the moment $\int \varphi f \mathrm{d}\mathbf{c} = \int_{-\infty}^{\infty}\int_{-\infty}^{\infty}\int_{-\infty}^{+\infty} \varphi f \mathrm{d}c_1 \mathrm{d}c_2 \mathrm{d}c_3$.

To derive the macroscopic equation model, setting $\varphi = m,\ mc_i,\ mc^2,\ mc^2 c_i, \ldots$ in equation (3), it is possible to construct an infinite hierarchy of moment equations [4, 6] that are truncated to the first eight moments:

$$\begin{aligned}
&\text{the mass density} && \rho = \int m f \mathrm{d}\mathbf{c},\\
&\text{the mass flux or the momentum density} && J_i = \int m c_i f \mathrm{d}\mathbf{c},\\
&\text{the energy density} && e = \tfrac{1}{2}\int m c^2 f \mathrm{d}\mathbf{c},\\
&\text{the heat flux} && q_i = \tfrac{1}{2}\int m c^2 c_i f \mathrm{d}\mathbf{c},
\end{aligned} \tag{4}$$

where the electric current density reads $S_i = -\frac{q}{m} J_i$.

From equation (3), the set of eight balance equations for the eight field variables ρ, J_i, e and q_i assumes the form

$$\begin{aligned}
&\frac{\partial \rho}{\partial t} + \frac{\partial J_k}{\partial x_k} = 0,\\
&\frac{\partial J_i}{\partial t} + \frac{\partial P_{ik}}{\partial x_k} - \left(-\tfrac{q}{m} E_i\right)\rho - \left(-\tfrac{q}{m}\epsilon_{ijk} B_k\right) J_j = G_i,\\
&\frac{\partial e}{\partial t} + \frac{\partial q_k}{\partial x_k} - \left(-\tfrac{q}{m} E_k\right) J_k = 0,\\
&\frac{\partial q_i}{\partial t} + \frac{1}{2}\frac{\partial \rho_{llik}}{\partial x_k} - \left(-\tfrac{q}{m} E_k\right)\left(P_{<ik>} + \tfrac{5}{3} e \delta_{ik}\right) - \left(-\tfrac{q}{m}\epsilon_{ijk} B_k\right) q_j = H_i,
\end{aligned} \tag{5}$$

if ϵ_{ijk} and δ_{ik} denote, as usual, the Levi-Civita tensor and the Kronecker tensor, $P_{ij} = \int m c_i c_j f \mathrm{d}\mathbf{c}$ represents the momentum flux, $\rho_{llik} = \int m c^2 c_i c_k f \mathrm{d}\mathbf{c}$ is the trace of the fourth moment, G_i and

H_i are the productions, while the square brackets stand for the traceless part of a symmetric tensor ($P_{ij} = P_{kk}\delta_{ij}/3 + P_{<ij>}$).

In order to obtain a closed set of field equations from system (5), the fluxes P_{ik} and ρ_{llik} and the productions G_i and H_i must be expressed in terms of the eight field variables ρ, J_i, e and q_i, by material dependent relations. Following the procedure of RET at macroscopic level, the closure of the system is obtained starting from the validity requirement of the entropy principle:

$$\frac{\partial h}{\partial t} + \frac{\partial \Phi_i}{\partial x_i} = \Sigma \geq 0, \tag{6}$$

for an entropy density h, an entropy flux Φ_i and an entropy production Σ. An approximation of the constitutive relations in the neighborhood of the equilibrium, the Gibbs relation and the Fermion statistics are the main ingredient used to reach the goal. In particular, we recall that at equilibrium the distribution function for fermion particles reads

$$f_{\mathrm{E}} = \frac{y}{e^{-\frac{m}{k_B}\frac{g}{T} + \frac{mc^2}{2k_B}\frac{1}{T}} + 1}, \tag{7}$$

if k_B is the Boltzmann constant, $g = \frac{e}{\rho} - \frac{Th_{\mathrm{E}}}{\rho} + \frac{p_{\mathrm{E}}}{\rho}$ is the specific free enthalpy, h_{E} denotes the entropy density at equilibrium, while p_{E} represents the equilibrium pressure, and $\frac{1}{y} = \frac{\mathsf{h}^3}{2m^3}$ [3, 4], where h denotes Plank's constant. To simplify the expressions of the final results, it is convenient to adopt as independent field variables the following ones: $\vartheta = 2k_BT/m$, J_i, $\alpha = -\frac{mg}{k_BT}$ and q_i. In this way one gets

$$\begin{aligned}
&P_{<ij>} = 0,\\
&e(\vartheta,\alpha) = 2\pi m y (\vartheta)^{\frac{5}{2}}\, i_4(\alpha),\\
&\rho(\vartheta,\alpha) = 4\pi m y (\vartheta)^{\frac{3}{2}}\, i_2(\alpha),\\
&a(\vartheta,\alpha) = \tfrac{4}{3}\pi m y (\vartheta)^{\frac{7}{2}}\, i_6(\alpha) + K(\vartheta),
\end{aligned} \tag{8}$$

for an arbitrary function $K(\vartheta)$ and the integral functions

$$i_n(\alpha) = \frac{I_n(\alpha)}{n+1} = \int_0^{+\infty} \frac{x^n}{e^{\alpha + x^2} + 1} dx. \tag{9}$$

The function $i_n(\alpha)$ is well-known in the framework of degenerate gas theory and several numerical techniques were adopted to calculate it, for an exhaustive review see [12]. It presents some peculiar properties, in particular we recall the following recurrence relation

$$\frac{\mathrm{d}i_n(\alpha)}{\mathrm{d}\alpha} = -\frac{n-1}{2} i_{n-2}(\alpha); \tag{10}$$

moreover, for completely degenerate gases (i.e. when $\alpha \to -\infty$) it holds $I_n(\alpha) \approx (-\alpha)^{(n+1)/2}$.

With regard to the production terms, the macroscopic approach requires that the production terms are expressed as a linear combination of J_i and q_i through a positive definite matrix that could depend on ϑ and α.

Through the comparison with Müller's kinetic approach it is possible to fix the remaining arbitrary functions: $K(\vartheta) = 0$ and

$$G_i = A_{11}(\vartheta,\alpha)J_i + A_{12}(\vartheta,\alpha)q_i, \qquad H_i = A_{21}(\vartheta,\alpha)J_i + A_{22}(\vartheta,\alpha)q_i, \tag{11}$$

where l is the mean free path of an electron between two collisions and

$$\begin{aligned}
A_{11}(\vartheta,\alpha) &= \frac{\vartheta^{1/2}}{l}\frac{I_4(\alpha)I_5(\alpha)-I_3(\alpha)I_6(\alpha)}{I_2(\alpha)I_6(\alpha)-I_4^2(\alpha)},\\
A_{12}(\vartheta,\alpha) &= \frac{2}{l\vartheta^{1/2}}\frac{I_3(\alpha)I_4(\alpha)-I_2(\alpha)I_5(\alpha)}{I_2(\alpha)I_6(\alpha)-I_4^2(\alpha)},\\
A_{21}(\vartheta,\alpha) &= \frac{\vartheta^{3/2}}{2l}\frac{I_4(\alpha)I_7(\alpha)-I_5(\alpha)I_6(\alpha)}{I_2(\alpha)I_6(\alpha)-I_4^2(\alpha)},\\
A_{22}(\vartheta,\alpha) &= \frac{\vartheta^{1/2}}{l}\frac{I_4(\alpha)I_5(\alpha)-I_2(\alpha)I_7(\alpha)}{I_2(\alpha)I_6(\alpha)-I_4^2(\alpha)}.
\end{aligned} \tag{12}$$

It was proven that the PDE system (5) is hyperbolic and equipped with an entropy law with concave entropy density in the neighborhood of equilibrium, if the electron gas is not completely degenerate [5].

2.1. The dimensionless variables

Here we will deal with processes independent of the time that will be described by the stationary equations

$$\begin{aligned}
\frac{\partial J_k}{\partial x_k} &= 0,\\
\frac{2}{3}\frac{\partial e}{\partial x_i} &= (-\frac{q}{m}E_i\rho - \frac{q}{m}\epsilon_{ijk}B_kJ_j) + A_{11}J_i + A_{12}q_i,\\
\frac{\partial q_k}{\partial x_k} &= -\frac{q}{m}E_kJ_k,\\
\frac{1}{2}\frac{\partial a}{\partial x_i} &= (-\frac{5q}{3m}E_ie - \frac{q}{m}\epsilon_{ijk}B_kq_j) + A_{21}J_i + A_{22}q_i.
\end{aligned} \tag{13}$$

In order to rewrite the system in dimensionless variables, we have to introduce preliminarly some suitable values of length, temperature and α denoted by D, T_0 and α_0, respectively.

Referring to relations (8) it is natural to propose the following reference quantities

$$\begin{aligned}
\vartheta_0 &= \frac{2kT_0}{m}, \quad \rho_0 = \frac{4\pi my\vartheta_0^{3/2}(-\alpha_0)^{3/2}}{3},\\
e_0 &= \frac{2\pi my\vartheta_0^{5/2}(-\alpha_0)^{5/2}}{5}, \quad a_0 = \frac{4}{3}\frac{\pi my\vartheta_0^{7/2}(-\alpha_0)^{7/2}}{7},
\end{aligned} \tag{14}$$

under the ussumption that $\alpha << -1$ and to define the following dimensionless quantities

$$\begin{aligned}
&\hat{x}_k = \frac{x_k}{D}, \quad \hat{\imath}_n(\alpha) = \frac{(n+1)i_n(\alpha)}{(-\alpha_0)^{(n+1)/2}}, \quad \hat{T} = \frac{T}{T_0}, \quad \hat{\vartheta} = \frac{\vartheta}{\vartheta_0},\\
&\hat{v}_k = \frac{v_k}{\sqrt{\vartheta_0}}, \quad \hat{J}_k = \frac{J_k}{\rho_0\sqrt{\vartheta_0}}, \quad \hat{q}_k = \frac{q_k}{\rho_0\vartheta_0^{3/2}}, \quad \hat{\alpha} = \frac{\alpha}{\alpha_0},\\
&\hat{\rho} = \frac{\rho}{\rho_0} = \hat{\vartheta}^{3/2}\hat{\imath}_2(\alpha), \quad \hat{e} = \frac{e}{e_0} = \hat{\vartheta}^{5/2}\hat{\imath}_4(\alpha), \quad \hat{a} = \frac{a}{a_0} = \hat{\vartheta}^{7/2}\hat{\imath}_6(\alpha).
\end{aligned} \tag{15}$$

Moreover, taking into account the previous relations, we will use the following dimensionless

IOP Conf. Series: Journal of Physics: Conf. Series **1035** (2018) 012003 doi:10.1088/1742-6596/1035/1/012003

coefficients

$$\hat{A}_{11}(\alpha) = \frac{\hat{\imath}_4(\alpha)\hat{\imath}_5(\alpha) - \hat{\imath}_3(\alpha)\hat{\imath}_6(\alpha)}{\hat{\imath}_2(\alpha)\hat{\imath}_6(\alpha) - \hat{\imath}_4^2(\alpha)}, \quad \hat{A}_{12}(\alpha) = \frac{\hat{\imath}_3(\alpha)\hat{\imath}_4(\alpha) - \hat{\imath}_2(\alpha)\hat{\imath}_5(\alpha)}{\hat{\imath}_2(\alpha)\hat{\imath}_6(\alpha) - \hat{\imath}_4^2(\alpha)},$$

$$\hat{A}_{21}(\alpha) = \frac{\hat{\imath}_4(\alpha)\hat{\imath}_7(\alpha) - \hat{\imath}_5(\alpha)\hat{\imath}_6(\alpha)}{\hat{\imath}_2(\alpha)\hat{\imath}_6(\alpha) - \hat{\imath}_4^2(\alpha)}, \quad \hat{A}_{22}(\alpha) = \frac{\hat{\imath}_4(\alpha)\hat{\imath}_5(\alpha) - \hat{\imath}_2(\alpha)\hat{\imath}_7(\alpha)}{\hat{\imath}_2(\alpha)\hat{\imath}_6(\alpha) - \hat{\imath}_4^2(\alpha)}, \tag{16}$$

$$\hat{F}_k^e = \frac{DqE_k}{m\vartheta_0}, \quad \hat{F}_k^B = \frac{DqB_k}{m\vartheta_0^{1/2}}, \quad \beta = \frac{D}{l}.$$

Hence, the stationary equations can be rewritten as

$$\frac{\partial \hat{J}_k}{\partial \hat{x}_k} = 0,$$

$$\hat{\vartheta}^{3/2}\hat{\imath}_4(\alpha)\frac{\partial \hat{\vartheta}}{\partial \hat{x}_i} - \frac{\hat{\vartheta}^{5/2}\hat{\imath}_2(\alpha)}{-\alpha_0}\frac{\partial \alpha}{\partial \hat{x}_i} =$$
$$\left[-\frac{2\hat{F}_i^e}{(-\alpha_0)}\hat{\vartheta}^{3/2}\hat{\imath}_2(\alpha) - \frac{6}{(-\alpha_0)^2}\epsilon_{ijk}\hat{F}_k^B\hat{J}_j\right] + \beta\left[\frac{2\hat{A}_{11}}{(-\alpha_0)^{1/2}}\hat{\vartheta}^{1/2}\hat{J}_i + \frac{4\hat{A}_{12}}{(-\alpha_0)^{3/2}}\frac{\hat{q}_i}{\hat{\vartheta}^{1/2}}\right], \tag{17}$$

$$\frac{\partial \hat{q}_k}{\partial \hat{x}_k} = -\hat{F}_k^e\hat{J}_k,$$

$$\hat{\vartheta}^{5/2}\hat{\imath}_6(\alpha)\frac{\partial \hat{\vartheta}}{\partial \hat{x}_i} - \frac{\hat{\vartheta}^{7/2}\hat{\imath}_4(\alpha)}{-\alpha_0}\frac{\partial \alpha}{\partial \hat{x}_i} =$$
$$\left[-\frac{2\hat{F}_i^e}{(-\alpha_0)}\hat{\vartheta}^{5/2}\hat{\imath}_4(\alpha) - \frac{4}{(-\alpha_0)^2}\epsilon_{ijk}\hat{F}_k^B\hat{q}_j\right] + \beta\left[\frac{2\hat{A}_{21}}{(-\alpha_0)^{1/2}}\hat{\vartheta}^{3/2}\hat{J}_i + \frac{4\hat{A}_{22}}{(-\alpha_0)^{3/2}}\hat{\vartheta}^{1/2}\hat{q}_i\right].$$

2.2. The 2D stationary set of equations

In this paper we will consider a metallic plate whose thickness is so small with respect to other dimensions that we can neglect it and describe the electron behavior inside the plate as a 2D problem in a bounded domain. Moreover, we will assume the possible presence of a magnetic flux density orthogonal to the plate, but the absence of any electromotive intensity. Hence the stationary 2D system is

$$B_{11}\frac{\partial \hat{\vartheta}}{\partial \hat{x}_i} + B_{12}\frac{\partial \hat{\alpha}}{\partial \hat{x}_i} = \epsilon_{ij3}\Omega\hat{J}_j + C_{11}\hat{J}_i + C_{12}\hat{q}_i,$$

$$B_{21}\frac{\partial \hat{\vartheta}}{\partial \hat{x}_i} + B_{22}\frac{\partial \hat{\alpha}}{\partial \hat{x}_i} = \epsilon_{ij3}\Gamma\hat{q}_j + C_{21}\hat{J}_i + C_{22}\hat{q}_i, \tag{18}$$

$$\frac{\partial \hat{q}_k}{\partial \hat{x}_k} = 0,$$

IOP Conf. Series: Journal of Physics: Conf. Series **1035** (2018) 012003 doi:10.1088/1742-6596/1035/1/012003

for $i = 1, 2$ and if

$$\begin{gathered} B_{11} = \hat{\vartheta}^{3/2}\hat{\imath}_4(\alpha), \qquad B_{12} = -\hat{\vartheta}^{5/2}\hat{\imath}_2(\alpha), \qquad B_{21} = \hat{\vartheta}^{5/2}\hat{\imath}_6(\alpha), \\ B_{22} = -\hat{\vartheta}^{7/2}\hat{\imath}_4(\alpha), \qquad C_{11} = \frac{2\beta\hat{A}_{11}(\alpha)}{(-\alpha)^{1/2}}\hat{\vartheta}^{1/2}, \qquad C_{12} = \frac{4\beta\hat{A}_{12}(\alpha)}{(-\alpha)^{3/2}\hat{\vartheta}^{1/2}}, \\ C_{21} = \frac{2\beta\hat{A}_{21}(\alpha)}{(-\alpha)^{1/2}}\hat{\vartheta}^{3/2}, \qquad C_{22} = \frac{4\beta\hat{A}_{22}(\alpha)}{(-\alpha)^{3/2}}\hat{\vartheta}^{1/2}, \\ \Omega = -\frac{6}{(-\alpha)^2}\hat{F}_3^B, \qquad \Gamma = -\frac{4}{(-\alpha)^2}\hat{F}_3^B. \end{gathered} \tag{19}$$

Now, if we assume that there is no electric current, i.e. $J_i = 0$ and we rewrite in a suitable way equations (18) for $i = 1, 2$, we get the following system:

$$\begin{aligned} \frac{\partial\hat{\alpha}}{\partial\hat{x}_1} &= \nu\frac{\partial\hat{\vartheta}}{\partial\hat{x}_1} + \varsigma\frac{\partial\hat{\vartheta}}{\partial\hat{x}_2}, \\ \frac{\partial\hat{\alpha}}{\partial\hat{x}_2} &= -\varsigma\frac{\partial\hat{\vartheta}}{\partial\hat{x}_1} + \nu\frac{\partial\hat{\vartheta}}{\partial\hat{x}_2}, \\ \hat{q}_1 &= \mu\frac{\partial\hat{\vartheta}}{\partial\hat{x}_2} + \eta\frac{\partial\hat{\vartheta}}{\partial\hat{x}_2}, \\ \hat{q}_2 &= -\eta\frac{\partial\hat{\vartheta}}{\partial\hat{x}_1} + \mu\frac{\partial\hat{\vartheta}}{\partial\hat{x}_2}, \\ \frac{\partial\hat{q}_1}{\partial\hat{x}_1} &+ \frac{\partial\hat{q}_2}{\partial\hat{x}_2} = 0, \end{aligned} \tag{20}$$

where

$$\begin{aligned} \nu(\alpha,\hat{\vartheta}) &= \frac{(-B_{21}C_{12}+B_{11}C_{22})(B_{22}C_{12}-B_{12}C_{22})-B_{11}B_{12}\Gamma^2}{\Lambda}, \\ \varsigma(\alpha,\hat{\vartheta}) &= \frac{(-B_{12}B_{21}+B_{11}B_{22})C_{12}\Gamma}{\Lambda}, \\ \mu(\alpha,\hat{\vartheta}) &= \frac{(B_{12}B_{21}-B_{11}B_{22})(-B_{22}C_{12}+B_{12}C_{22})}{\Lambda}, \\ \eta(\alpha,\hat{\vartheta}) &= -\frac{(B_{12}B_{21}-B_{11}B_{22})B_{12}\Gamma}{\Lambda}, \\ \Lambda(\alpha,\hat{\vartheta}) &= (B_{22}C_{12}-B_{12}C_{22})^2 + (B_{12}\Gamma)^2. \end{aligned} \tag{21}$$

We underline that all the previous coefficients depend on B_3 through Γ and, in particular, ς and η vanish in the absence of a magnetic field. This fact will play a relevant role in the next sections.

2.3. The linearized equations

In order to simplify the integration of the previous equations, we linearize system (17) in the neighborhood of $\alpha = \alpha_0$ and $\vartheta = \vartheta_0$, so that, setting

$$\nu_0 = \nu(\alpha_0, \hat{\vartheta}_0), \quad \varsigma_0 = \nu(\alpha_0, \hat{\vartheta}_0), \quad \mu_0 = \mu(\alpha_0, \hat{\vartheta}_0), \quad \eta_0 = \eta(\alpha_0, \hat{\vartheta}_0), \tag{22}$$

IOP Conf. Series: Journal of Physics: Conf. Series **1035** (2018) 012003 doi:10.1088/1742-6596/1035/1/012003

system (20) reduces to

$$\frac{\partial\hat{\alpha}}{\partial\hat{x}_1} = \nu_0\frac{\partial\hat{\vartheta}}{\partial\hat{x}_1} + \varsigma_0\frac{\partial\hat{\vartheta}}{\partial\hat{x}_2},$$

$$\frac{\partial\hat{\alpha}}{\partial\hat{x}_2} = -\varsigma_0\frac{\partial\hat{\vartheta}}{\partial\hat{x}_1} + \nu_0\frac{\partial\hat{\vartheta}}{\partial\hat{x}_2}, \tag{23}$$

$$\nabla^2\hat{\vartheta} = 0,$$

and is equipped with the following algebraic relations:

$$\hat{q}_1 = \mu_0\frac{\partial\hat{\vartheta}}{\partial\hat{x}_2} + \eta_0\frac{\partial\hat{\vartheta}}{\partial\hat{x}_2},$$

$$\hat{q}_2 = -\eta_0\frac{\partial\hat{\vartheta}}{\partial\hat{x}_1} + \mu_0\frac{\partial\hat{\vartheta}}{\partial\hat{x}_2}. \tag{24}$$

For a problem in a bounded domain described by the previous set of equations it is necessary to prescribe the values of the temperature or of the heat flux along the boundaries (to solve the Laplace equation). In addition, a value of α has to be fixed in a point of the boundary in order to determine completely the α quantity. We imagine, for example, to set $\alpha = \alpha_0$ in a vertex of the plate.

3. Heat transfer problems

We consider the application of the previous model to heat transfer problems. For a classical gas the stationary heat transfer problem in a bounded domain is described equivalently assigning a heat flux or different temperatures on the boundaries (see for example [13, 14, 15]). What we will verify here and in the following section is that for an electron gas the two assignments give rise to different phenomena. To present both cases, let us imagine to deal with a square plate whose side legth is D, and introduce a reference system with the axis passing through two consecutive sides of the plate: in this way, we can assume that $\hat{x}_1 \in [0,1]$, $\hat{x}_2 \in [0,1]$. Referring to reasonable values for the mean free path and for the mean electron density, accounting for a small sample at room temperature and a very high maximum magnetic flux density, we will consider here and in the next section the following parameter values: $\beta \leq 10^6$, $\alpha_0 = -100$, $|\hat{F}_3^B| \leq 10^5$. The value of α_0 will be prescribed in the vertex $(\hat{x}_1, \hat{x}_2) = (0,0)$.

3.1. Temperature assignment

First of all, let us study the assignment of the temperature values (or, it is the same, of $\hat{\vartheta}$) on the boundaries of the plate. Just to analyse a simple case (although a bit rough) we prescibe the same temperature values on three sides of the plate and a different temperature on the remaining one. The behavior of the temperature corresponding to this case can be described analytically, referring to the classical solution of the Laplace equations, under the previous conditions.

Concerning the temperature, let T_0 be the temperature common to three sides $x_1 = 0$, $x_2 = 0$ and $x_2 = 1$ so that $\hat{\vartheta} = \hat{\vartheta}_0 = 1$ on these boundaries, while on $x_1 = 1$, $\hat{\vartheta} = \hat{\vartheta}_1$. Then, the solution for $\hat{\vartheta}$ reads:

$$\hat{\vartheta}(\hat{x}_1, \hat{x}_2) = \hat{\vartheta}_0 + \Delta\hat{\vartheta}\sum_{n=1}^{\infty} c_n \sinh(n\pi\hat{x}_1)\sin(n\pi\hat{x}_2), \tag{25}$$

while the expression for $\hat{\alpha}$ is

$$\hat{\alpha}(\hat{x}_1, \hat{x}_2) = K_0 + \Delta\hat{\vartheta}\sum_{n=1}^{\infty} c_n \left[\nu \sinh(n\pi\hat{x}_1)\sin(n\pi\hat{x}_2) + \varsigma\cosh(n\pi\hat{x}_1)\cos(n\pi\hat{x}_2),\right] \tag{26}$$

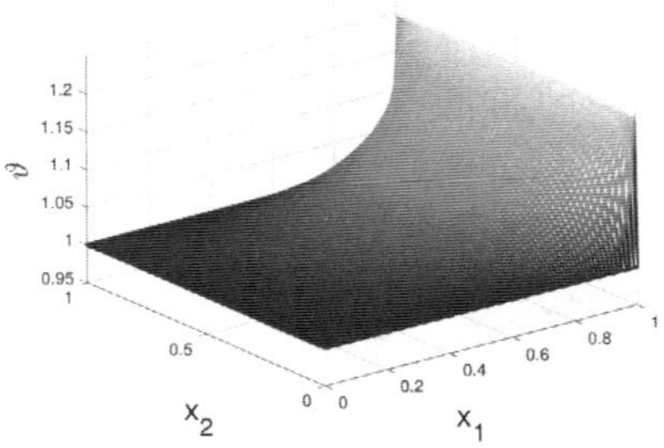

Figure 1. A 3D-plot of the analytical $\hat{\vartheta}$ solution ($\hat{\vartheta}_1 = 1.2$, $\hat{F}_3^B = 10^5$, $\beta = 10^6$).

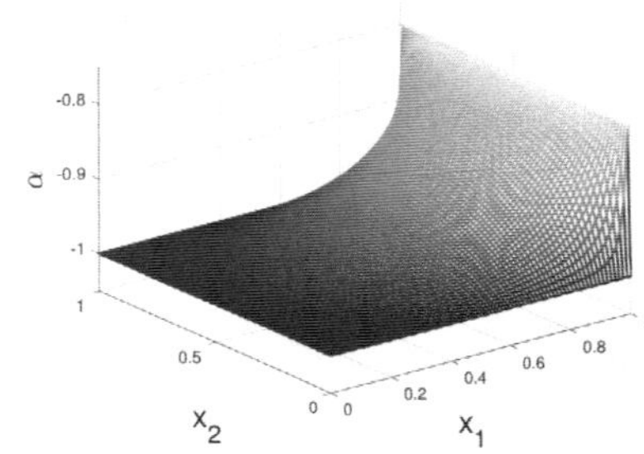

Figure 2. A 3D-plot of the analytical $\hat{\alpha}$ solution ($\hat{\vartheta}_1 = 1.2$, $\hat{F}_3^B = 10^5$, $\beta = 10^6$).

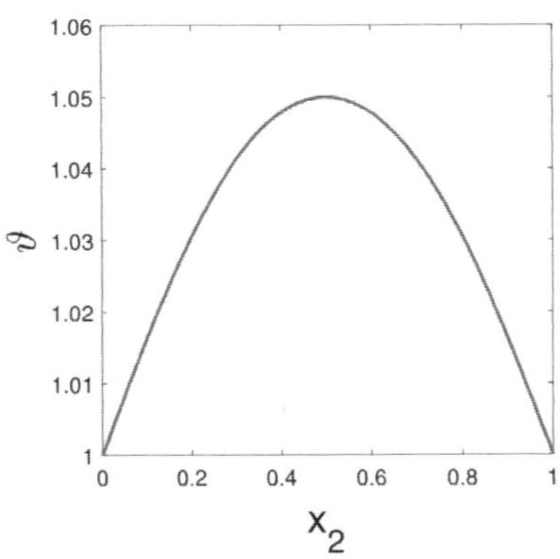

Figure 3. The transversal profile of the analytical expression of $\hat{\vartheta}$ independent of $\hat{F}_3^B$ ($\hat{\vartheta}_1 = 1.2$, $\beta = 10^6$).

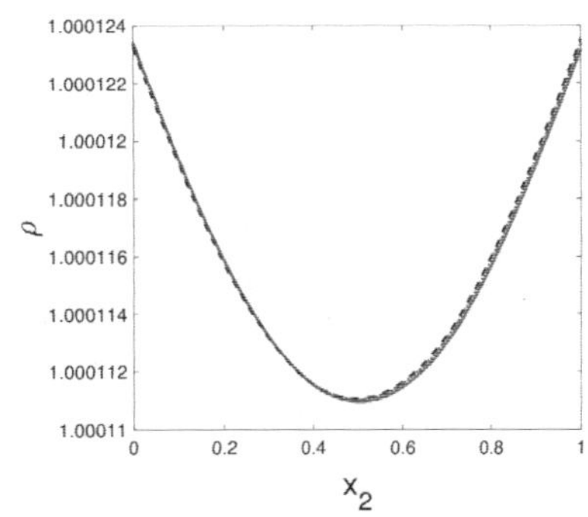

Figure 4. The transversal profile of the analytical expression of $\hat{\rho}$ for $\hat{F}_3^B = -10^5$ (continuous line), $\hat{F}_3^B = 10^5$ (dashed line), $\hat{F}_3^B = 0$ (dotted line), ($\hat{\vartheta}_1 = 1.2$, $\beta = 10^6$).

with

$$c_n = \frac{2[1-(-1)^n]}{n\pi \sinh(n\pi)}, \quad \Delta\hat{\vartheta} = \hat{\vartheta}_1 - \hat{\vartheta}_0, \quad K_0 = \hat{\alpha}_0 - \varsigma\Delta\hat{\vartheta}\sum_{n=1}^{\infty} c_n. \tag{27}$$

Unfortunately, due to the discontinuities in $(\hat{x}_1, \hat{x}_2) = (1, 0)$ and $(\hat{x}_1, \hat{x}_2) = (1, 1)$, the previous series are only pointwise and not uniformly convergent; anyway, in the present case it is possible to get an analytical expression of the solution.

Figures 1 and 2 show a 3D-plot of $\hat{\vartheta}$ and $\hat{\alpha}$ as functions of the two spatial variables. It is evident that in the present case the temperature behaviour does not depend on the magnetic flux density. In contrast, α and also the electron mass density are related to B_3 through the coefficient μ and η. Nevertheless, very small differences are observable for $\hat{\alpha}$ and $\hat{\rho}$, when the value of $\hat{F}_3^B$ is varied. This fact is evident in Figures 3 and 4, where the profile of $\hat{\vartheta}$ and $\hat{\rho}$ along the $\hat{x}_2$ axis ($\hat{x}_2 = 0.5$) is presented for different values of $\hat{F}_3^B$. We remark that $\hat{\rho}$ was defined as dimensionless variable before the linearization, but here it was determined through a linear approximation, so in $(\hat{x}_1, \hat{x}_2) = (0, 0)$ in general $\hat{\rho} \neq 1$.

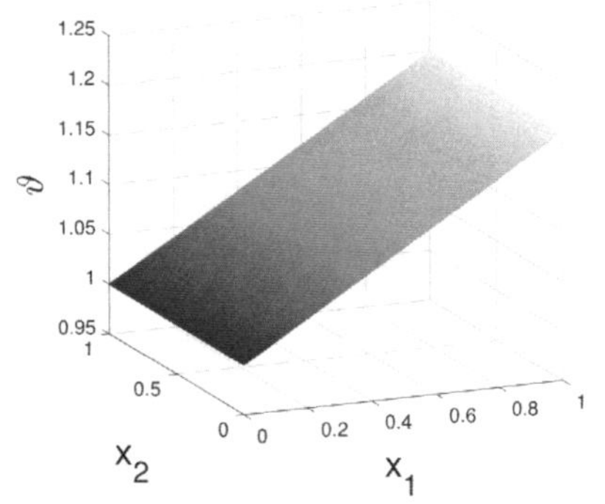

Figure 5. A 3D-plot of the $\hat{\vartheta}$ behaviour, in the case of Section 3.2 ($\hat{\vartheta}_1 = 1.2$, $\hat{F}_3^B = 10^5$, $\beta = 10^6$).

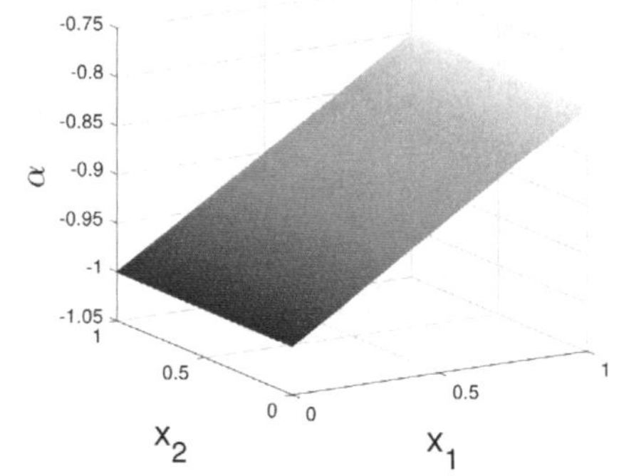

Figure 6. A 3D-plot of the $\hat{\alpha}$ behaviour, in the case of Section 3.2 ($\hat{\vartheta}_1 = 1.2$, $\hat{F}_3^B = 10^5$, $\beta = 10^6$).

3.2. Heat flux assignment and Righi-Leduc effect

The Righi-Leduc effect [16, 17] is a well-known effect in the literature. In the case of a metallic plate it predicts a transversal temperature gradient ($\frac{\partial \hat{\vartheta}}{\partial \hat{x}_2} \neq 0$) in the presence of a longitudinal heat flux q_1 and an orthogonal magnetic field B_3 under the adiabatic condition $q_2 = 0$. This fact is easily verified for the previous equations (20). If we impose that $\hat{q}_2 = 0$ we get

$$\begin{aligned} \frac{\partial \hat{\vartheta}}{\partial \hat{x}_2} &= \frac{\eta}{\mu}\frac{\partial \hat{\vartheta}}{\partial \hat{x}_1} = -\frac{B_{12}\Gamma}{(-B_{22}C_{12} + B_{12}C_{22})}\frac{\partial \hat{\vartheta}}{\partial \hat{x}_1}, \\ \hat{q}_1 &= \frac{\mu^2 + \eta^2}{\mu}\frac{\partial \hat{\vartheta}}{\partial \hat{x}_1} = \frac{(B_{12}B_{21} - B_{11}B_{22})}{(-B_{22}C_{12} + B_{12}C_{22})}\frac{\partial \hat{\vartheta}}{\partial \hat{x}_1}. \end{aligned} \tag{28}$$

The coefficient for the Righi-Leduc effect is η/μ and through a mathematical technique similar to the one used in the Appendices of [5], it is possible to show that if the electron gas is not completely degenerate, it holds

$$sgn(\eta/\mu) = -sgn(B_3), \tag{29}$$

in agreement with previous studies [1, 2].

In the framework of a real experiment, in a metallic plate it is not possible to impose that $\hat{q}_2 = 0$ everywhere. What is reasonable to prescribe at the boundaries is a temperature difference at two parallel sides and the vanishing orthogonal component of the heat flux at the other sides ($\hat{q}_2 = 0$). Unlike [5], we neglect here the effect due to the interaction with the boundaries for two reasons: first of all we are interested to isolate the thermomagnetic phenomenon, secondly we are dealing here with a rough linearized model and the interaction effects could represent a more refined aspect. The previous conditions bring to a problem that cannot be easily solved analitycally, even if we refer here to the linearized equations (23). The results obtained through a numerical integration of (23), based on finite difference schemes and carried on with Matlab©, are shown in the following Figures. In particular, Figure 5 and 6 show the profile of $\hat{\vartheta}$ and $\hat{\alpha}$ as functions of the spatial variables.

The details about the longitudinal and the transverse tendency of the field variables could not be observed through a 3D-plot. Therefore, in Figure 7 we present the $\hat{\vartheta}$ profile for $\hat{x}_1 = 0.5$. In Figure 8 we resort again to a 3D-picture to get an overview of the temperature difference for different values of the magnetic field.

IOP Conf. Series: Journal of Physics: Conf. Series **1035** (2018) 012003 doi:10.1088/1742-6596/1035/1/012003

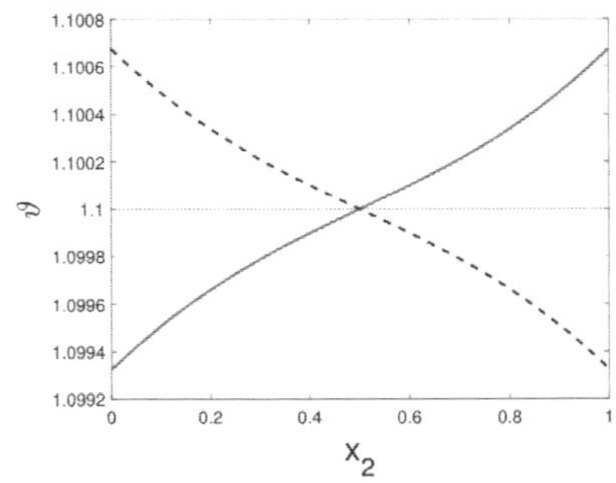

Figure 7. $\hat{\vartheta}$ as a function of $\hat{x}_2$ ($\hat{\vartheta}_1 = 1.2$, $\beta = 10^6$) for different values of $\hat{F}_3^B$: $\hat{F}_3^B = -10^5$ continuous line, $\hat{F}_3^B = 10^5$ dashed line, $\hat{F}_3^B = 0$ dotted green line.

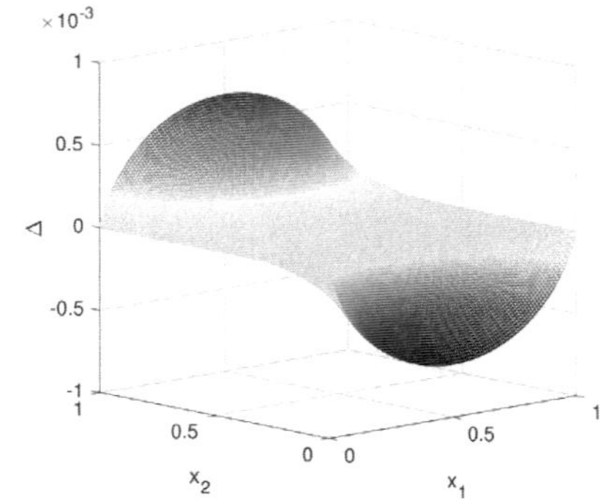

Figure 8. A 3D-plot of the difference $\Delta = \hat{\vartheta}|_{\hat{F}_3^B=-10^5} - \hat{\vartheta}|_{\hat{F}_3^B=0}$ ($\hat{\vartheta}_1 = 1.2$, $\beta = 10^6$).

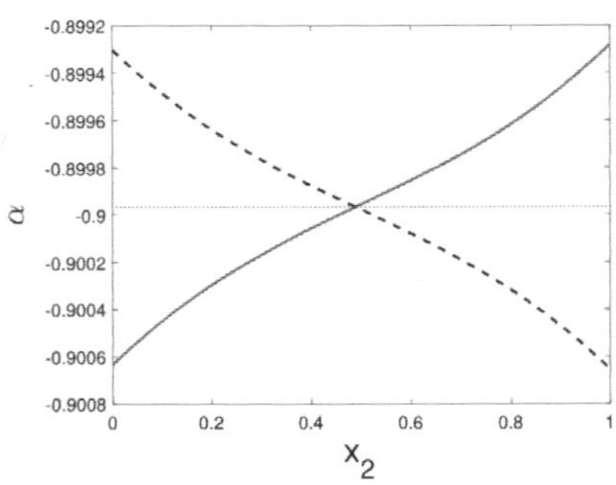

Figure 9. The behaviour of $\hat{\alpha}$ as a function of $\hat{x}_2$ ($\hat{\vartheta}_1 = 1.2$, $\beta = 10^6$) for different values of $\hat{F}_3^B$: $\hat{F}_3^B = -10^5$ continuous line, $\hat{F}_3^B = 10^5$ dashed line, $\hat{F}_3^B = 0$ dotted green line.

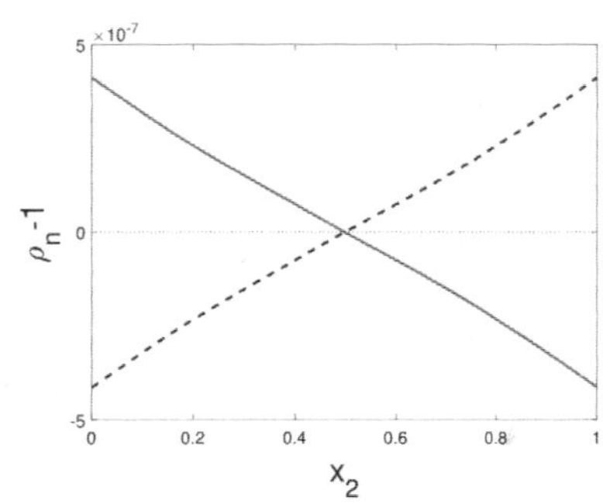

Figure 10. The behaviour of $\hat{\rho}_n = \rho(x_1, x_2)/\rho(0.5, 0.5)$ as a function of $\hat{x}_2$ ($\hat{\vartheta}_1 = 1.2$, $\beta = 10^6$) for different values of $\hat{F}_3^B$: $\hat{F}_3^B = -10^5$ continuous line, $\hat{F}_3^B = 10^5$ dashed line, $\hat{F}_3^B = 0$ dotted green line.

The transversal behavior of $\hat{\alpha}$ and $\hat{\rho}$ along the $\hat{x}_2$-axis (when $\hat{x}_1 = 0.5$) is illustrated in Figures 9 and 10. To better compare the different solutions of $\hat{\rho}$ we have plotted the variable normalized with respect to the central value. Figure 11 presents a comparison of the transversal profile of $\hat{\vartheta}$ for different values of β, that is to say for different mean free paths (i.e. different metals). For Figure 12 we introduce the quantity $\vartheta_n(\hat{x}_1, \hat{x}_2) = \hat{\vartheta}(\hat{x}_1, \hat{x}_2)/\hat{\vartheta}(0.5, 0.5)$ in order to compare the transversal profile for different values of $\Delta\hat{\vartheta}$. From the previous analysis it is natural to conclude that the Righi-Leduc effect is enhanced for processes further from equilibrium: if the magnetic field, the difference between the temperature at the boundaries or the mean free path increases, the same happens also to the transversal temperature gradient.

These results confirm the idea that the choice of the boundary conditions in a heat transfer problem plays a crucial role to describe different phenomena related to metal electrons.

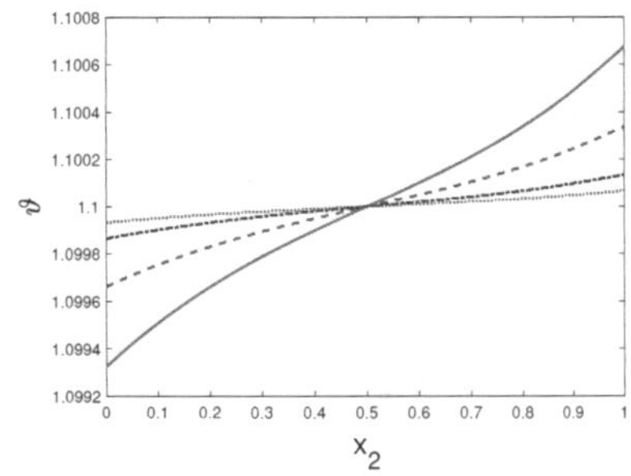

Figure 11. The profile of $\hat{\vartheta}$ as a function of $\hat{x}_2$ when $\hat{x}_1 = 0.5$ ($\hat{\vartheta}_1 = 1.2$, $\hat{F}_3^B = -10^5$) for different values of β: $\beta = 10^6$ continuous line, $\beta = 2 \times 10^6$ dashed line, $\beta = 5 \times 10^6$ dotted-dashed purple line, $\beta = 10^7$ dotted green line.

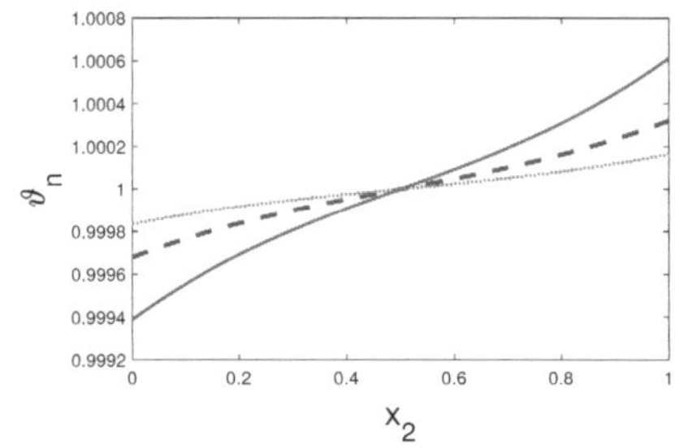

Figure 12. The profile of $\hat{\vartheta}_n = \theta(x_1, x_2)/\theta(0.5, 0.5)$ as a function of $\hat{x}_2$ when $\hat{x}_1 = 0.5$ ($\beta = 10^6$, $\hat{F}_3^B = -10^5$) for different values of $\Delta\hat{\vartheta}$: $\Delta\hat{\vartheta} = 0.2$ continuous line, $\Delta\hat{\vartheta} = 0.1$ dashed line, $\Delta\hat{\vartheta} = 0.05$ dotted green line.

4. Conclusions

This paper aims to a preliminary study of 2D heat transfer problem for metal electrons confined in a very thin plate. The role of the magnetic field is also investigated. Although the oversemplification of the equation system (few moments and linearization), it is possible to describe qualitatively well the main features of the phenomena and in particular to "capture" the Righi-Leduc effect, obtaining results also for the electron density behavior. To our knowledge, these cases are described here through a 2D integration for the first time in the literature. The linearized model allows a semi-analytical approach to the solution and constitute the first step to a more general treatment of the problem, that is already in progress.

Acknowledgments

This paper is dedicated with deep esteem to Masaru Sugiyama. It was supported by GNFM-INdAM.

References

[1] Sommerfeld A 1952 *Vorlesungen über theoretische Physik Thermodynamik und Statistik* (Wiesbaden: Dietrichshe Verlagbuchhandlung)
[2] Sommerfeld A and Bethe H 1956 *Handbuch der Physik. XIV* (Heidelberg: Springer)
[3] Müller I 1976 *Acta Mechanica* **24** 117
[4] Müller I and Ruggeri T 1998 *Rational extended thermodynamics* (New York: Springer)
[5] Barbera E and Brini F 2018 *Ricerche Mat.* (*Preprint* `https://doi.org/10.1007/s11587-018-0377-9`)
[6] Ruggeri T and Sugiyama M 2015 *Rational extended thermodynamics beyond the monatomic gas* (New York: Springer)
[7] Arima T, Taniguchi S, Ruggeri T and Sugiyama M 2012 *Continuum Mech. Thermodyn.* **24** 271
[8] Pennisi S and Ruggeri T 2017 *Annals of Physics* **377** 414
[9] Trovato M 2014 *Acta Applicandae Mathematicae* **132** 605
[10] Taniguchi S, Arima T, Ruggeri T and Sugiyama M 2014 *Phys. Rev. E* **89** 013025
[11] Arima T, Taniguchi S, Ruggeri T and Sugiyama M 2013 *Continuum Mech. Thermodyn.* **25** 727
[12] Ruggeri T and Trovato M 2004 *Continuum Mech. Thermodyn.* **16** 551
[13] Müller I and Ruggeri T 2004 *J. Non-newtonian Fluid Mech.* **119** 139
[14] Barbera E, Brini F and Sugiyama M 2014 *Acta Applicandae Mathematicae* **132** 41
[15] Arima T, Barbera E, Brini F and Sugiyama M 2014 *Physics Letters A* **378** 2695
[16] Leduc S 1887 *C.R. Hebd. Sèances Acad. Sci.* **104** 1783
[17] Righi A 1887 *C.R. Hebd. Sèances Acad. Sci.* **105** 168

IOP Conf. Series: Journal of Physics: Conf. Series **1035** (2018) 012004 doi:10.1088/1742-6596/1035/1/012004

On the deviation of statistical diffusion-exponent fluctuation from Poisson-like distribution

Yuichi Itto

Science Division, Center for General Education, Aichi Institute of Technology, Aichi 470-0392, Japan

E-mail: itto@aitech.ac.jp

Abstract. The deviation of the statistical distribution of the diffusion-exponent fluctuations from a Poisson-like distribution is discussed for virus in cytoplasm of a living cell. The deviation is shown to obey the multivariate Gaussian distribution for a class of small deviations. A brief discussion is also made about the behavior of such a deviation from a geometric viewpoint.

1. Introduction

An exotic diffusion has experimentally been observed in the infection pathway of adeno-associated viruses in living HeLa cells [1,2]. The virus fluorescently labeled with a dye molecule exhibits stochastic motions in cytoplasm of the cell in two different forms, which are the free form and the form being contained in the endosome. Analysis of each trajectory of such viruses offers the mean square displacement denoted here as

$$\overline{x^2} \sim t^{\alpha}, \tag{1}$$

where t is elapsed time and α is termed the diffusion exponent. The diffusion property obtained from the experimental results is as follows [1]: normal diffusion with $\alpha = 1$, whereas the case with $0 < \alpha < 1$, which is referred to as anomalous diffusion [3,4]. A fact to be stressed is that α fluctuates between 0.5 and 0.9, which depends on localized areas of the cytoplasm. This presents an evidence for the existence of the diffusion-exponent fluctuations over the cytoplasm and, thus, makes the diffusion observed outstanding in anomalous diffusion under vital investigation in the literature. The fluctuations may reflect heterogeneous structure of the cytoplasm [1].

In a recent work [5], an attempt has been made to clarify the statistical property of the fluctuations mentioned above based on the experimental data. There, the following Poisson-like distribution of the fluctuations, which is purposely expressed here in the case of discrete values of α, has been proposed:

$$P_{\alpha_i} \sim e^{\lambda \alpha_i} \quad (i = 1, 2, \ldots, A), \tag{2}$$

where A is the total number of different values of the exponent, α_i the ith value of the exponent and λ a positive constant. A basic premise for this distribution is that the time scale of variation of the fluctuations is much larger than that of stochastic motion of the virus in the localized area. That is, α slowly varies on a long time scale, but is assumed to be approximately constant.

If this assumption is relaxed, then the fluctuation distribution to be observed may deviate from the distribution in equation (2) only slightly. Here is a naive question: is it possible to determine the behavior of such a deviation?

IOP Conf. Series: Journal of Physics: Conf. Series **1035** (2018) 012004 doi:10.1088/1742-6596/1035/1/012004

In this article, we examine deviation of the fluctuation distribution from the Poisson-like distribution, which is henceforth simply referred to as deviation. In particular, we focus our attention on the behavior of small deviation. Such a deviation is found to explicitly obey the multivariate Gaussian distribution [6]. In addition, we briefly discuss the behavior of the deviation from a geometric viewpoint.

2. Entropy and its maximization

To develop our discussion, we introduce the entropy associated with the diffusion-exponent fluctuations, from which the Poisson-like distribution in equation (2) can be derived [5]. Such an entropy turns out to enable us to examine the deviation.

Suppose that the cytoplasm plays a role of a medium for stochastic motions of the virus in the two different forms, which is divided into many small virtual blocks. Each block is identified with the localized area of the cytoplasm, and the exponent in equation (1) fluctuates depending on these local blocks accordingly. The exponent is assumed to be approximately constant due to the slowness mentioned above. (This assumption will be relaxed later.)

The entropy is introduced as follows. There is no information about how the exponent locally distributes over the cytoplasm, and this is the situation we are considering here. Let us construct all of possible collections of the blocks, each of which corresponds to the medium and is distinguishable from each other with respect to the local fluctuations. Clearly, each collection is equivalent to the medium at the statistical level of the fluctuations. Denoting the total number of these collections by G, the entropy is introduced as

$$S = \frac{\ln G}{N} \tag{3}$$

with N being the total number of blocks in the medium. This gives a measure of uncertainty about the fluctuations. It is necessary to evaluate G, since otherwise equation (3) is formal. To do so, recall that the exponent in equation (1) is determined by average over each trajectory. It is therefore considered that the trajectory in a given localized area is included in the average, whereas the trajectories in other localized areas are not. This may imply that the local blocks are statistically independent from each other with respect to the exponent. Accordingly, for the medium with a set of exponents, $\{\alpha_i\}_{i=1,2,\ldots,A}$, G is given by $G = N!/\prod_{i=1}^{A} n_{\alpha_i}!$, where n_{α_i} is the number of blocks with α_i in the medium. It is clear that $\sum_{i=1}^{A} n_{\alpha_i} = N$ is satisfied. Now, largeness of the number of blocks in the medium has been discussed in Ref. [6] by evaluating the volume of the block. The discussion given there suggests that n_{α_i}'s are indeed large enough, allowing us to write S in equation (3) as the following form of the Shannon entropy:

$$S \simeq S[P] = -\sum_{i=1}^{A} P_{\alpha_i} \ln P_{\alpha_i}, \tag{4}$$

where $P_{\alpha_i} = n_{\alpha_i}/N$ is the probability of finding α_i in a given block of the medium. This symbol for the probability should not be mixed still at this stage with that for the distribution in equation (2).

Now, the Poisson-like distribution can be derived based on the entropy as shown in Ref. [5], where only information is supposed to be available about the statistical property of the fluctuations. In fact, with the constraints on the average of the exponent, $\sum_{i=1}^{A} \alpha_i P_{\alpha_i} = \overline{\alpha}$, as well as the normalization condition, $\sum_{i=1}^{A} P_{\alpha_i} = 1$, the maximum entropy principle [7] for $S[P]$ in equation (4) turns out to lead to the distribution in equation (2) as the stationary solution of the variational problem with respect to P_{α_i}, in which λ plays a role of a positive Lagrange multiplier for the constraint on the average.

3. Deviation of the fluctuation distribution

Up to the above stage, we have assumed that the exponent is approximately constant. In this section, we examine the deviation of the statistical fluctuation by relaxing this assumption. Our idea for it is based on Einstein's theory of fluctuations of the thermodynamic quantities [8].

IOP Conf. Series: Journal of Physics: Conf. Series **1035** (2018) 012004 doi:10.1088/1742-6596/1035/1/012004

A basic observation is as follows. Let G_{max} be the maximum value of G, for which the entropy in equation (3) takes the maximum value given by $S_{\text{max}} = (\ln G_{\text{max}})/N$. This situation is realized for the Poisson-like distribution in equation (2). Now, suppose that G is decreased to a certain value given by $G' = G_{\text{max}}/e^k$, where k is a positive constant. For the entropy, this corresponds to $S' = (\ln G')/N = S_{\text{max}} - k/N$. Accordingly, S' is approximately equal to S_{max}, if k/N is negligible compared to S_{max}, (recalling largeness of the total number of blocks, N), implying that the statistical fluctuation associated with G' can be realized.

In what follows, we consider that the statistical fluctuation to be realized can deviate slightly from the Poisson-like fluctuation in equation (2) due to its slow variation. Then, from the above observation, we assume that such a statistical fluctuation is described by the fluctuation distribution associated with a certain value of G, which is smaller than G_{max}. Keeping this in mind, let us consider a set of n_{α_i}'s under the same constraints imposed for the set associated with G_{max}. All of possible sets are distinct in the sense that the values of n_{α_i}'s in a given set are different from those in other sets. Then, we describe the probability of finding the medium in the state with the statistical fluctuation concerning with the nth set by

$$W^{(n)} = \frac{G^{(n)}}{\sum_m G^{(m)}}, \tag{5}$$

where $G^{(n)}$ stands for the value of G for the nth set.

We here notice the following point. $W^{(n)}$ can be expressed in terms of the entropy for the nth set, $S^{(n)} = (\ln G^{(n)})/N$, as

$$W^{(n)} \propto e^{NS^{(n)}}. \tag{6}$$

This has a similarity with Einstein's theory of fluctuations [8-10], i.e., the reversal of Boltzmann's relation: $W \propto e^S$, where S is the thermodynamic entropy and the Boltzmann constant is set equal to unity. [Note that S appearing here is not the entropy in equation (3).]

Below, we evaluate the behavior of the deviation based on this similarity. Let $P_{\alpha_i}^{(n)}$ be the fluctuation distribution associated with the nth set, which is connected to P_{α_i} in equation (2):

$$P_{\alpha_i}^{(n)} = P_{\alpha_i} + \Delta P_{\alpha_i}^{(n)} \quad (i = 1, 2, \ldots, A), \tag{7}$$

where $\Delta P_{\alpha_i}^{(n)}$ denotes the deviation of the fluctuation distribution from the Poisson-like distribution and is supposed to be small. As mentioned above, the distribution in equation (7) should be examined under the constraints not only on the normalization, $\sum_{i=1}^{A} P_{\alpha_i}^{(n)} = 1$, but also on the average of the exponent, $\sum_{i=1}^{A} \alpha_i P_{\alpha_i}^{(n)} = \overline{\alpha}$, which require the deviation, $\{\Delta P_{\alpha_i}^{(n)}\}_{i=1,2,\ldots,A}$, to fulfill the following two conditions

$$\sum_{i=1}^{A} \Delta P_{\alpha_i}^{(n)} = 0, \qquad \sum_{i=1}^{A} \alpha_i \Delta P_{\alpha_i}^{(n)} = 0, \tag{8}$$

respectively. (It is noted that the present approach is applicable for $A > 2$, which seems to hold [1], otherwise the deviation identically vanishes.)

From equation (4), expanding $S[P^{(n)}]$ up to the second order of $\Delta P_{\alpha_i}^{(n)}$'s, $S^{(n)}$ is calculated to be

$$S^{(n)} \simeq S_{\text{max}} - \frac{1}{2} \sum_{i=1}^{A} \frac{1}{P_{\alpha_i}} \Delta P_{\alpha_i}^{(n)2}, \tag{9}$$

where the conditions in equation (8) have been used. Substitution of equation (9) into equation (6) immediately reads

$$W^{(n)} \sim W_{\text{max}} \exp\left[-\frac{N}{2} \sum_{i=1}^{A} \frac{1}{P_{\alpha_i}} \Delta P_{\alpha_i}^{(n)2} \right], \tag{10}$$

IOP Conf. Series: Journal of Physics: Conf. Series **1035** (2018) 012004 doi:10.1088/1742-6596/1035/1/012004

where $W_{\max} \equiv G_{\max}/\sum_m G^{(m)}$ is the probability of finding the medium in the state with the Poisson-like fluctuation. Now, it is obvious from equation (8) that for instance, each of $\Delta P^{(n)}_{\alpha_{A-1}}$ and $\Delta P^{(n)}_{\alpha_A}$ can be written in terms of the deviation, $\{\Delta P^{(n)}_{\alpha_i}\}_{i=1,2,\ldots,A-2}$. Taking this into account, $W^{(n)}$ is also found to have the following form:

$$W^{(n)} \sim W_{\max} \exp\left[-\frac{N}{2}\sum_{i,j=1}^{A-2} h_{ij}\Delta P^{(n)}_{\alpha_i}\Delta P^{(n)}_{\alpha_j}\right], \tag{11}$$

where $H = (h_{ij})$ is the positive-definite symmetric matrix with the following elements:

$$h_{ij} = \frac{1}{P_{\alpha_i}}\delta_{ij} + \frac{1}{P_{\alpha_{A-1}}}\frac{(\alpha_i - \alpha_A)(\alpha_j - \alpha_A)}{(\alpha_{A-1} - \alpha_A)^2} + \frac{1}{P_{\alpha_A}}\frac{(\alpha_i - \alpha_{A-1})(\alpha_j - \alpha_{A-1})}{(\alpha_{A-1} - \alpha_A)^2}. \tag{12}$$

Thus, from equation (10) [or, equivalently (11)], we see that the deviation explicitly obeys the multivariate Gaussian distribution.

4. Deviation: A geometric viewpoint

The above results seem to give information about the scale of deviation and allow us to discuss the behavior of such a deviation from a geometric viewpoint.

To see this, we here suppose that such a scale is obtained through the extension of $W^{(n)}$ in equation (10) with respect to the deviation such as its half-width:

$$\sum_{i=1}^{A} \frac{1}{P_{\alpha_i}}\Delta P^{(n)*2}_{\alpha_i} = \frac{2\ln 2}{N}, \tag{13}$$

where $\Delta P^{(n)*}_{\alpha_i}$ describes a typical scale of deviation for the nth set. Clearly, equation (13) can be rewritten as follows:

$$\frac{\Delta P^{(n)*2}_{\alpha_1}}{a_1{}^2} + \frac{\Delta P^{(n)*2}_{\alpha_2}}{a_2{}^2} + \cdots + \frac{\Delta P^{(n)*2}_{\alpha_A}}{a_A{}^2} = 1 \tag{14}$$

with $a_i = \sqrt{(2\ln 2)P_{\alpha_i}/N}$ $(i = 1, 2, \ldots, A)$. Therefore, regarding $\{\Delta P^{(n)}_{\alpha_i}\}_{i=1,2,\ldots,A}$ as a coordinate point in the A-dimensional space of the deviations, equation (14) shows that under the conditions in equation (8), the deviation, $\{\Delta P^{(n)*}_{\alpha_i}\}_{i=1,2,\ldots,A}$, lies on an A-dimensional ellipsoid centered at the origin in the space. In other words, in this space, the conditions in equation (8) yield two different hyperplanes: one for $\Delta P^{(n)}_{\alpha_1} + \Delta P^{(n)}_{\alpha_2} + \cdots + \Delta P^{(n)}_{\alpha_A} = 0$, and the other for $\alpha_1\Delta P^{(n)}_{\alpha_1} + \alpha_2\Delta P^{(n)}_{\alpha_2} + \cdots + \alpha_A\Delta P^{(n)}_{\alpha_A} = 0$, each of which passes through the origin and intersects with each other. A set of points of the intersection then meets the above ellipsoid, on which the deviation lies. This may contribute to understanding a geometric interpretation of the behavior of the deviation.

5. Concluding remarks

We have discussed the deviation of the statistical distribution of the diffusion-exponent fluctuations for adeno-associated virus in cytoplasm of a HeLa cell. Examining the entropy associated with such fluctuations in the spirit of Einstein's theory of fluctuations, we have shown that the small deviation from the Poisson-like fluctuation is governed by the multivariate Gaussian distribution. We have also mentioned the behavior of such a deviation in view of geometry.

It may be worth pointing out that a kinetic theory for diffusion of the virus has been presented in Ref. [5] (see Refs. [11,12] for a theoretical framework deriving it in a consistent manner) based on the Poisson-like fluctuation. (One might care about the continuum limit of entropy given there since a careful treatment in measure theory is required for such a limit [7]. This point has been examined in a recent work [13].) In addition, it has been shown [14] that a fluctuation distribution observed in an experiment for virus capsid in a cell nucleus can be derived by the entropic approach. Thus, it is of interest to develop the present discussions for these studies.

Acknowledgement
The present article is based on the author's talk at the Meeting on "Nonequilibrium thermodynamics and statistical physics: From rational modeling to its applications" (16-17 March, 2017, Fukuoka, Japan). He would like to thank the organizers of the meeting for providing him with opportunity to give a talk.

References

[1] Seisenberger G, Ried M U, Endreß T, Büning H, Hallek M and Bräuchle C 2001 *Science* **294** 1929
[2] Bräuchle C, Seisenberger G, Endreß T, Ried M U, Büning H and Hallek M 2002 *ChemPhysChem* **3** 299
[3] Bouchaud J -P and Georges A 1990 *Phys. Rep.* **195** 127
[4] Metzler R, Jeon J -H, Cherstvy A G and Barkai E 2014 *Phys. Chem. Chem. Phys.* **16** 24128
[5] Itto Y 2012 *J. Biol. Phys.* **38** 673
[6] Itto Y 2016 *Physica* A **462** 522
[7] Rosenkrantz R D (ed.) 1989 *E. T. Jaynes: Papers on Probability, Statistics and Statistical Physics* (Dordrecht: Kluwer)
[8] Einstein A 1910 *Ann. Phys.* **33** 1275
[9] Landau L D and Lifshitz E M 1980 *Statistical Physics* 3rd ed. (Oxford: Pergamon Press)
[10] Pais A 1982 *'Subtle is the Lord . . .'* (Oxford: Oxford University Press)
[11] Itto Y 2014 *Phys. Lett.* A **378** 3037
[12] Itto Y 2017 *Frontiers in Anti-Infective Drug Discovery* vol 5, ed Atta-ur-Rahman and M Iqbal Choudhary (Sharjah: Bentham Science Publishers) p 3
[13] Itto Y 2018 *Open Conf. Proc. J.* **9** 1
[14] Itto Y 2018 *Phys. Lett.* A **382** 1238

IOP Conf. Series: Journal of Physics: Conf. Series **1035** (2018) 012005 doi:10.1088/1742-6596/1035/1/012005

A New BGK Model for Relativistic Kinetic Theory of Monatomic and Polyatomic Gases

Sebastiano Pennisi[1], **Tommaso Ruggeri**[2]

[1] Department of Mathematics and Informatics, University of Cagliari, Cagliari, Italy
[2] Department of Mathematics and Alma Mater, Research Center on Applied Mathematics AM [2], University of Bologna, Bologna, Italy

E-mail: `spennisi@unica.it`, `tommaso.ruggeri@unibo.it`

Abstract. We propose a possible generalization of the BGK collisional term in the relativistic kinetic theory and we compare our model with previous ones. The present model has the advantage in that it satisfies the conservation of particle number, energy-momentum and the H-theorem in the Eckart frame without constraints on particular local equilibrium state. In the last part of the paper we extend the model to include also the polyatomic gas.

1. Introduction

In the kinetic theory one of most complex questions is to give an explicit expression for the collisional term. In the classical case the simplest case for the collisional term is the one proposed in the well-known paper by Bhatnagar-Gross-Krook [1] that is known as BGK model. The BGK model is simple, but has the main ingredients needed to satisfy a collisional term, i.e. the conservation of mass, momentum and energy and the H-theorem. Of course the model is oversimplified because only one relaxation time is present in the associated production terms related to the moments, and the phenomenological coefficients obtained via Chapman-Enskog procedure as heat conductivity, and viscosity are not so good in comparison with experiments. Nevertheless BGK model has been largely used not only in Boltzmann equation (see e.g. [2]), but also in the context of Rational Extended Thermodynamics (ET) to determine the production terms of the differential system [3]. In particular, this can be said if we take into account the recent progress obtained in ET to include polyatomic gas in the theory [4].

In the relativistic framework the BGK model was reconsidered first by Marle [5] and successively a different proposal was given by Anderson and Witting (AW) [6] . In this paper we discuss first the previous proposal and then we will propose, both for monatomic and for polyatomic gases, a variant of AW model proving the conservation of particles number, momentum-energy and the H-Theorem in the Eckart frame instead of the Landau-Lifshitz one.

The results presented here can be useful to evaluate the production terms in the relativistic ET of monatomic gas [7] and for polyatomic gas [8].

2. The relativistic Boltzmann Chernikov equation

Let us consider, as starting point, the Boltzmann Chernikov equation for the determination of the distribution function $f(x^\alpha, p^\beta)$:

$$p^\alpha \partial_\alpha f = Q \,, \tag{1}$$

IOP Conf. Series: Journal of Physics: Conf. Series **1035** (2018) 012005 doi:10.1088/1742-6596/1035/1/012005

where x^α is the space-time coordinate, p^α is the four-momentum ($p_\alpha p^\alpha = m^2c^2$), m is the particle mass, c is the light velocity $\partial_\alpha = \partial/\partial x^\alpha$, and Q is the collisional term ($\alpha = 0, 1, 2, 3$). Repeated indexes, as usual, denote summation from 0 to 3. There is now the problem of determining the production term Q appearing in this equation. There are different possible models in literature for Q (see, e.g. [2, 9]). In any way a physical Q must vanish in any local equilibrium state and moreover, it must guarantee the conservation of particle number and momentum-energy and need to satisfy the H-theorem.

It is well known that from (1), we can define the moment equations. In particular the following first five ones:

$$\partial_\alpha V^\alpha = P, \qquad \partial_\alpha T^{\alpha\beta} = P^\beta, \tag{2}$$

with

$$V^\alpha = mc \int_{\Re^3} f p^\alpha d\mathbf{P}, \qquad T^{\alpha\beta} = c \int_{\Re^3} f p^\alpha p^\beta d\mathbf{P}, \tag{3}$$

and

$$P = mc \int_{\Re^3} Q \, d\mathbf{P}, \qquad P^\alpha = c \int_{\Re^3} Q p^\alpha d\mathbf{P}, \tag{4}$$

where $d\mathbf{P} = dp^1 dp^2 dp^3/p^0$. As V^α is the particle-particle flux and $T^{\alpha\beta}$ is the energy-momentum tensor, we need that (2) must be conservation laws and therefore we need that the collisional term Q must satisfy the conditions $P = 0, P^\alpha = 0$, i.e.

$$\int_{\Re^3} Q \, d\mathbf{P} = 0, \qquad \int_{\Re^3} Q p^\alpha d\mathbf{P} = 0, \qquad \alpha = 0, 1, 2, 3. \tag{5}$$

Moreover defining the entropy four vector h^α and the entropy production Σ:

$$h^\alpha = -k_B c \int_{\Re^3} f \ln f \, p^\alpha d\mathbf{P}, \qquad \Sigma = -k_B c \int_{\Re^3} Q \ln f \, d\mathbf{P}, \tag{6}$$

(k_B is the Boltzmann constant), then from (1) we obtain the entropy law:

$$\partial_\alpha h^\alpha = \Sigma. \tag{7}$$

The H-theorem requires that the entropy production must be non-negative ($\Sigma \geq 0$) and therefore we need to require that Q satisfies also the inequality

$$\int_{\Re^3} Q \ln f \, d\mathbf{P} \leq 0. \tag{8}$$

To sum up, we have that any physical collisional term Q must satisfy the five equations (5) and the inequality (8).

3. Eckart and Landau-Lifshitz frames

We recall that in the relativistic hydrodynamics there are two possible choices of the 4-velocity. The most popular one is the Eckart 4-velocity and the other one is the Landau-Lifshitz 4-velocity (see e.g. [9, 10]).

IOP Conf. Series: Journal of Physics: Conf. Series **1035** (2018) 012005 doi:10.1088/1742-6596/1035/1/012005

– The *Eckart frame* or *particle frame* corresponds to the frame in which there is no dissipative contribution to the rest-mass density current and to the energy density. In this case the four velocity U^β is collinear with V^β and we have:

$$V^\alpha = V^\alpha_E, \qquad \left(T^{\alpha\beta} - T^{\alpha\beta}_E\right) U_\beta = 0, \tag{9}$$

where the index E indicates a generic local equilibrium state.

In the Eckart frame it is possible to rewrite V^α and $T^{\alpha\beta}$ in terms of the usual physical quantities:

$$\begin{aligned} V^\alpha &= \rho U^\alpha \\ T^{\alpha\beta} &= \sigma^{\langle\alpha\beta\rangle} + (p+\Pi)h^{\alpha\beta} + \frac{1}{c^2}(q^\alpha U^\beta + q^\beta U^\alpha) + \frac{e}{c^2}U^\alpha U^\beta, \end{aligned} \tag{10}$$

where U^α is the Eckart four-velocity ($U^\alpha U_\alpha = c^2$), $\rho = nm = c^{-1}\sqrt{V^\alpha V_\alpha}$ is the number density, p is the pressure, Π is the dynamical pressure, $h^{\alpha\beta}$ is the projector tensor $h^{\alpha\beta} = -g^{\alpha\beta} + \frac{1}{c^2}U^\alpha U^\beta$ with $g^{\alpha\beta} = diag(1, -1, -1, -1)$ being the metric tensor, the symbol $\langle\cdots\rangle$ denotes the 3-dimensional traceless part of a tensor, $\sigma^{\langle\alpha\beta\rangle} = T^{\mu\nu}\left(h^\alpha_\mu h^\beta_\nu - \frac{1}{3}h^{\alpha\beta}h_{\mu\nu}\right)$ is the viscous deviatoric stress, e is the energy, and $q^\alpha = -h^\alpha_\mu U_\nu T^{\mu\nu}$ is the heat flux. From the previous definitions we have $h^{\alpha\beta}U_\beta = 0, q^\alpha U_\alpha = 0, \sigma^{\langle\alpha\beta\rangle}U_\beta = 0$ and therefore only 14 fields are independent.

– The *Landau-Lifshitz frame* or *energy frame* represents the frame in which there is no net energy flux, i.e., it is the frame defined by the conditions

$$(V^\alpha - V^\alpha_E)U_{L\alpha} = 0, \qquad \left(T^{\alpha\beta} - T^{\alpha\beta}_E\right) U_{L\alpha}U_{L\beta} = 0, \tag{11}$$

where U^α_L indicates the Landau-Lifshitz four-velocity.

In the Landau-Lifshitz frame V^α and $T^{\alpha\beta}$ have the following decompositions:

$$\begin{aligned} V^\alpha &= \rho\left(U^\alpha_L - \frac{1}{p+e}q^\alpha\right) \\ T^{\alpha\beta} &= \sigma^{\langle\alpha\beta\rangle} + (p+\Pi)h^{\alpha\beta}_L + \frac{e}{c^2}U^\alpha_L U^\beta_L, \end{aligned} \tag{12}$$

with projector $h^{\alpha\beta}_L = -g^{\alpha\beta} + \frac{1}{c^2}U^\alpha_L U^\beta_L$. The expressions (10) and (12) are equivalent if the following relation between Eckart and Landau-Lifshitz 4-velocities holds:

$$U^\alpha_L = U^\alpha + \frac{1}{e+p}q^\alpha,$$

and if we neglect second order terms in the non-equilibrium variables $\Pi, q^\alpha, \sigma^{\langle\alpha\beta\rangle}$.

4. The relativistic BGK approximation

In the relativistic framework the most important generalization of BGK approximation was made by Marle [5] and successively by Anderson and Witting [6].

4.1. The Marle BGK model

The Marle model [5] is an extension of the non-relativistic BGK model in the Eckart frame:

$$Q = -\frac{m}{\tau}\left(f - f_E\right), \tag{13}$$

where τ is the relaxation time in the rest frame where the momentum of particles is zero, and $f_E = f_J$ is the Jüttner equilibrium distribution (see [2,8]):

$$f_J = \exp\left(\xi - \frac{1}{k_B T} U_\beta p^\beta\right), \quad \xi = -1 + \frac{m\, g_r}{k_B\, T}. \tag{14}$$

In (14) g_r denotes the relativistic chemical potential

$$g_r = \frac{e+p}{\rho} - TS.$$

T and S are respectively the temperature and the entropy density ($\rho S = h^\alpha u_\alpha$). We notice that g_r is in reality the chemical potential except for c^2 term. In fact taking into account that the energy is composed of two parts: Part due to the internal energy ε and part to the rest-energy:

$$e = \rho(\varepsilon + c^2),$$

we have

$$g_r = g + c^2, \qquad g = \varepsilon + \frac{p}{\rho} - TS\,,$$

where g denotes the usual chemical potential.

In the Marle BGK approximation, by inserting (13), the conditions (5) and the inequality (8) become, respectively:

$$\int_{\Re^3} (f - f_E) d\mathbf{P} = 0, \qquad \int_{\Re^3} (f - f_E) p^\alpha d\mathbf{P} = 0, \tag{15}$$

$$\int_{\Re^3} (f - f_E) \ln f d\mathbf{P} \geq 0. \tag{16}$$

The second of (15) is equivalent to the first condition of the Eckart frame definition (9) and the inequality (16) holds for analogous arguments of classical BGK (see e.g. [2]). The most weak condition is $(15)_1$ that becomes a constraint on the distribution function f. In fact the five conditions (15) can be interpreted as the equations for the five unknown particular local equilibrium variables T, g_r and U_α appearing in the Jüttner equilibrium distribution function (14) and these coefficients in reality depend themselves on f. The existence of the solution of this problem was proved in [11]. Another weak point of the Marle model is that the relaxation time becomes unbounded in the case of particles with zero rest mass [2].

4.2. The Anderson and Witting BGK model

The Anderson-Witting model provides another expression of the Q, described in the Landau-Lifshitz frame and has been widely used:

$$Q = -\frac{U_{L\mu} p^\mu}{c^2 \tau} (f - f_E)\,, \tag{17}$$

where $U_{L\mu}$ indicates the four-velocity according with the Landau-Lifshitz definition. In the AW model the conditions that guarantee the conservation laws (5) and the H-theorem (8) become respectively:

$$U_{L\mu} \int_{\Re^3} (f - f_E)\, p^\mu\, d\mathbf{P} = 0, \qquad U_{L\mu} \int_{\Re^3} (f - f_E) p^\mu\, p^\alpha\, d\mathbf{P} = 0\,, \tag{18}$$

IOP Conf. Series: Journal of Physics: Conf. Series **1035** (2018) 012005 doi:10.1088/1742-6596/1035/1/012005

$$U_{L\mu} \int_{\Re^3} (f - f_E) \ln f \, p^\mu \, d\mathbf{P} \geq 0. \tag{19}$$

The conditions (18) are coincident with (11) that define the Landau-Lifshitz frame and the inequality (19) is satisfied (see [2] for the proof). Therefore the AW model satisfies all the requirements necessary for the collisional term and moreover in the classical limit converges to the classical BGK model. The weak point is that the frame is the Landau-Lifshitz frame that is less used in literature also for the complexity in the conservation of number of particle [10]. A comparison between the two BGK models and also on the Marle-Grad 14 moments was studied in [12]. In particular in [12], the Cauchy problem has been studied for the linearized kinetic equations with the Marle and Anderson-Witting models, and compared the resulting dispersion relations with the 14-moment theory.

5. A new relativistic BGK approximation for monatomic gases

The question arises if it is possible to construct a relativistic BGK in the Eckart frame without the problematic present in the Marle model. We notice that, if we substitute in the AW model the 4-velocity of Eckart the condition $(18)_2$ is not anymore satisfied because the left side is not zero but proportional to the heat flux 4-vector q^α. We want to prove in this paper that it is possible to construct a sort of generalization of the AW model that satisfies automatically the definition of the Eckart frame (9), satisfies the H-theorem, and reduces to the BGK in the classical limit.

Before introducing it, let us consider the following equilibrium moment of third order:

$$A_E^{\alpha\beta\mu} = \frac{c}{m} \int_{\Re^3} f_E p^\alpha p^\beta p^\mu d\mathbf{P} = a U^\alpha U^\beta U^\mu + b \left(h^{\alpha\beta} U^\mu + h^{\alpha\mu} U^\beta + h^{\beta\mu} U^\alpha \right). \tag{20}$$

Inserting the Jüttner distribution into (20), it is simple to verify that the coefficients appearing in (20) are

$$a = \rho \left(3 \frac{K_3(\gamma)}{\gamma K_2(\gamma)} + 1 \right), \qquad b = k_B n T \frac{K_3(\gamma)}{K_2(\gamma)} = p \frac{K_3(\gamma)}{K_2(\gamma)},$$

with

$$\gamma = \frac{mc^2}{k_B T}, \tag{21}$$

and $K_n(\gamma)$ denotes the modified Bessel functions of second kind.

We propose as variant of Relativistic BGK this form of the collisional term:

$$Q = \frac{U_\alpha p^\alpha}{c^2 \tau} \left(f_E - f - f_E \, p^\mu q_\mu \frac{1}{bmc^2} \right). \tag{22}$$

Comparing these expression with (17), we see that we have now the extra term $-f_E \, p^\mu q_\mu \frac{1}{bmc^2}$ inside the parenthesis and the use of Eckart 4-velocity instead of the Landau-Lifshitz one. We can prove

Theorem 1 (Conservation Laws) *The collisional term* (22) *conserves the number of particle and the energy momentum in the Eckart frame i.e.* $P = P^\alpha = 0$.

IOP Conf. Series: Journal of Physics: Conf. Series **1035** (2018) 012005 doi:10.1088/1742-6596/1035/1/012005

The proof is immediate. Inserting (22) in (4) and taking into account the Eckart frame definition (9) we have:

$$
\begin{aligned}
P =& \frac{m}{c\tau} U_\alpha \int_{\Re^3} \left(f_E - f - f_E p^\mu q_\mu \frac{1}{bmc^2} \right) p^\alpha d\mathbf{P} = \\
=& - \frac{1}{b\tau c^4} U_\alpha T_E^{\alpha\mu} q_\mu = - \frac{e}{b\tau c^4} U^\mu q_\mu = 0, \\
P^\beta =& \frac{1}{c\tau} U_\alpha \int_{\Re^3} \left(f_E - f - f_E \, p^\mu q_\mu \frac{1}{bmc^2} \right) p^\alpha p^\beta d\mathbf{P} = \\
=& \frac{1}{c^2\tau} U_\alpha \left(T_E^{\alpha\beta} - T^{\alpha\beta} - \frac{1}{bc^2} A_E^{\alpha\beta\mu} q_\mu \right) = - \frac{1}{c^2\tau} (q^\beta - q^\beta) = 0.
\end{aligned}
\tag{23}
$$

Theorem 2 (H-Theorem) *The collisional term (22) satisfies the H-theorem up to second order terms with respect to non-equilibrium variables.*

Proof- For our collisional term (22) we have from $(6)_2$:

$$
\Sigma = - \frac{k_B}{c\tau} U_\alpha \int_{\Re^3} \left(f_E - f - f_E \, p^\mu q_\mu \frac{1}{bmc^2} \right) \ln f \, p^\alpha d\mathbf{P}. \tag{24}
$$

Let us consider the right hand side of this relation, but with $\ln f_E$ instead of $\ln f$, that is,

$$
- \frac{k_B}{c\tau} U_\alpha \int_{\Re^3} \left(f_E - f - f_E \, p^\mu q_\mu \frac{1}{bmc^2} \right) \ln f_E \, p^\alpha d\mathbf{P}. \tag{25}
$$

Thanks to eq. (14) it becomes

$$
- \frac{k_B}{c\tau} U_\alpha \int_{\Re^3} \left(f_E - f - f_E \, p^\mu q_\mu \frac{1}{bmc^2} \right) \left[-1 + \frac{m g_r}{k_B T} - \frac{1}{k_B T} U_\beta p^\beta \right] p^\alpha d\mathbf{P}
$$

which is zero, thanks to (23). Consequently, we can subtract (25) from (24) so that it becomes

$$
\Sigma = - \frac{k_B}{c\tau} U_\alpha \int_{\Re^3} \left(f_E - f - f_E \, p^\mu q_\mu \frac{1}{bmc^2} \right) \ln \frac{f}{f_E} p^\alpha d\mathbf{P}.
$$

Now we sum to this expression, $(23)_2$ contracted by $-\frac{k_B}{bmc^3} q_\beta$; so it becomes

$$
\Sigma = \frac{k_B}{c\tau} U_\alpha \int_{\Re^3} \left(-f_E + f + f_E \, p^\mu q_\mu \frac{1}{bmc^2} \right) \left(\ln \frac{f}{f_E} + p^\beta q_\beta \frac{1}{bmc^2} \right) p^\alpha d\mathbf{P}.
$$

Now, Taylor's expansion of $\ln \frac{f}{f_E}$ around $f = f_E$, dropped at the first order expansion, is

$$
\ln \frac{f}{f_E} = \ln \frac{f_E}{f_E} + \frac{1}{f_E} (f - f_E) + O(2) \tag{26}
$$

so that we have

$$
\Sigma = \frac{k_B}{c\tau} U_\alpha \int_{\Re^3} \frac{1}{f_E} \left(-f_E + f + f_E \, p^\mu q_\mu \frac{1}{bmc^2} \right)^2 p^\alpha d\mathbf{P} + O(3)
$$

from which, taking into account that $U_\alpha p^\alpha > 0$, it follows $\Sigma \geq 0$.

IOP Conf. Series: Journal of Physics: Conf. Series **1035** (2018) 012005 doi:10.1088/1742-6596/1035/1/012005

5.1. *Classical Limit of* (22)

Taking into account that the time and space components of U^α and of p^α are

$$U^\alpha \equiv \bar{\Gamma}(c, v^i), \quad p^\alpha \equiv m\Gamma(c, \xi^i), \quad \bar{\Gamma} = \frac{1}{\sqrt{1-\frac{v^2}{c^2}}}, \quad \Gamma = \frac{1}{\sqrt{1-\frac{\xi^2}{c^2}}}$$

(v^i and ξ^i are respectively the macroscopic and the microscopic velocity), and evaluating (22) in the rest frame $v^i = 0$, we have:

$$Q = \frac{m\Gamma}{\tau}\left(f_E - f + f_E \frac{1}{bc^2}\Gamma q_i \xi^i\right).$$

Taking into account that [8]

$$\lim_{c\to\infty} \Gamma = 1, \quad \lim_{c\to\infty} b = p, \quad \lim_{c\to\infty} f_E = \frac{1}{m^3} f_M, \quad \lim_{c\to\infty} f = \frac{1}{m^3} f^C,$$

where f_M is the Maxwellian and f^C denote the classical distribution function solution of

$$\partial_t f^C + \xi^i \partial_i f^C = Q^C, \quad x^0 = ct$$

we obtain:

$$Q^C = \lim_{c\to\infty} \frac{m^2 Q}{\Gamma} = \frac{1}{\tau}(f_M - f^C).$$

Therefore our BGK relativistic variant converges to the classical BGK!

6. A Relativistic BGK approximation for polyatomic gases

In [8] starting from the classical ideas for polyatomic gases introduced in [4, 13–17] we proposed a generalized Boltzmann-Chernikov equation that has the same form of (1) but the extended distribution function $f \equiv f(x^\alpha, p^\beta, \mathcal{I})$ depends on an extra variable $\mathcal{I}$ that takes into account the energy due to the internal degrees of freedom of a molecule. We consider instead of (3), the following moments:

$$\begin{aligned} V^\alpha &= mc \int_{\Re^3}\int_0^{+\infty} f p^\alpha \phi(\mathcal{I})\, d\mathbf{P}\, d\mathcal{I}, \\ T^{\alpha\beta} &= \frac{1}{mc}\int_{\Re^3}\int_0^{+\infty} f\left(mc^2 + \mathcal{I}\right) p^\alpha p^\beta\, \phi(\mathcal{I})\, d\mathbf{P}\, d\mathcal{I}. \end{aligned} \tag{27}$$

The meaning of $(27)_2$ is that the energy and the momentum in relativity are components of the same tensor and we expect that, besides the energy at rest, there is a contribution due to the internal structure, as in the case of a classical polyatomic gas. $\phi(\mathcal{I})$ is the state density of the internal mode, that is, $\phi(\mathcal{I})\, d\mathcal{I}$ represents the number of the internal states of a molecule having the internal energy between $\mathcal{I}$ and $\mathcal{I} + d\mathcal{I}$.

The macroscopic internal energy

$$\varepsilon = \frac{e}{mn} - c^2,$$

in the classical limit, when γ given by (21) tend to infinity, converges to the one of a classical polyatomic gas [8]:

$$\lim_{\gamma\to\infty} \varepsilon = \frac{D}{2}\frac{k_B}{m}T,$$

provided that the measure

$$\phi(\mathcal{I}) = \mathcal{I}^a,$$

where the constant is given by

$$a = \frac{D-5}{2},$$

and $D = 3 + f^i$ is related to the degrees of freedom of a molecule given by the sum of the space dimension 3 for the translational motion and the contribution from the internal degrees of freedom $f^i \geq 0$ related to the rotation and vibration. For monatomic gases $D = 3$ and $a = -1$.

As in the monatomic case we introduce the following equilibrium triple tensor

$$\begin{aligned} A_E^{\alpha\beta\mu} &= \frac{c}{m}\int_{\Re^3}\int_0^{+\infty} f_E p^\alpha p^\beta p^\mu \left(1+\frac{\mathcal{I}}{mc^2}\right)^2 \phi(\mathcal{I}) d\mathbf{P}\, d\mathcal{I} = \\ &= aU^\alpha U^\beta U^\mu + b\left(h^{\alpha\beta}U^\mu + h^{\alpha\mu}U^\beta + h^{\beta\mu}U^\alpha\right), \end{aligned} \tag{28}$$

where f_E is the generalized Jüttner equilibrium distribution function obtained in [8] via *Maximum Entropy Principle*:

$$f_E = \exp\left[\xi - \left(1+\frac{\mathcal{I}}{mc^2}\right)\frac{1}{k_B T}U_\beta p^\beta\right], \quad \xi = -1 + \frac{m\, g_r}{k_B\, T}, \tag{29}$$

The expression of the coefficients appearing in $(28)_3$ is

$$\begin{aligned} a &= \frac{\rho}{\gamma A(\gamma) K_2(\gamma)}\int_0^{+\infty}\left[3K_3(\gamma^*) + \gamma^* K_2(\gamma^*)\right]\phi(\mathcal{I})\, d\mathcal{I}, \\ b &= \frac{nk_B T}{A(\gamma)K_2(\gamma)}\int_0^{+\infty} K_3(\gamma^*)\phi(\mathcal{I})\, d\mathcal{I}, \\ A(\gamma) &= \frac{\gamma}{K_2(\gamma)}\int_0^{+\infty}\frac{K_2(\gamma^*)}{\gamma^*}\phi(\mathcal{I})\, d\mathcal{I}, \end{aligned}$$

where we put

$$\gamma^* = \gamma\left(1+\frac{\mathcal{I}}{mc^2}\right).$$

We consider as a variant of Relativistic Polyatomic BGK this form of the collisional term:

$$Q = \frac{U_\alpha p^\alpha}{c^2 \tau}\left(f_E - f - f_E\, p^\mu q_\mu \frac{1+\frac{\mathcal{I}}{mc^2}}{bmc^2}\right). \tag{30}$$

We want to prove the following:

Theorem 3 (Conservation Laws in Polyatomic Gas) *The collisional term (30) conserves the number of particle and the energy momentum, i.e.*

$$\begin{aligned} P &= \partial_\alpha V^\alpha = mc\int_{\Re^3}\int_0^{+\infty} Q\,\phi(\mathcal{I})\, d\mathbf{P}\, d\mathcal{I} = 0, \\ P^\beta &= \partial_\alpha T^{\alpha\beta} = c\int_{\Re^3}\int_0^{+\infty} Q\, p^\beta\left(1+\frac{\mathcal{I}}{mc^2}\right)\phi(\mathcal{I}) d\mathbf{P}\, d\mathcal{I} = 0. \end{aligned} \tag{31}$$

IOP Conf. Series: Journal of Physics: Conf. Series **1035** (2018) 012005 doi:10.1088/1742-6596/1035/1/012005

The proof is immediate. Inserting (30) in (31) and taking into account the Eckart frame (9) we have:

$$
\begin{aligned}
P =& \frac{m}{c\tau} U_\alpha \int_{\Re^3} \int_0^{+\infty} \left(f_E - f - f_E p^\mu q_\mu \frac{1 + \frac{\mathcal{I}}{mc^2}}{bmc^2} \right) p^\alpha \, \phi(\mathcal{I}) d\mathbf{P} \, d\mathcal{I} = \\
=& -\frac{1}{b\tau c^4} U_\alpha T_E^{\alpha\mu} q_\mu = -\frac{e}{b\tau c^4} U^\mu q_\mu = 0, \\
P^\beta =& \frac{1}{c\tau} U_\alpha \int_{\Re^3} \int_0^{+\infty} \left(f_E - f - f_E \, p^\mu q_\mu \frac{1 + \frac{\mathcal{I}}{mc^2}}{bmc^2} \right) p^\alpha p^\beta \cdot \\
& \cdot \left(1 + \frac{\mathcal{I}}{mc^2} \right) \phi(\mathcal{I}) d\mathbf{P} \, d\mathcal{I} = \\
=& \frac{1}{c^2 \tau} U_\alpha \left(T_E^{\alpha\beta} - T^{\alpha\beta} - \frac{1}{bc^2} A_E^{\alpha\beta\mu} q_\mu \right) = -\frac{1}{c^2\tau} (q^\beta - q^\beta) = 0.
\end{aligned}
\tag{32}
$$

Theorem 4 (H-Theorem in Polyatomic Gas) *Let*

$$
h^\alpha = -k_B \, c \int_{\Re^3} \int_0^{+\infty} f \ln f \, p^\alpha \phi(\mathcal{I}) d\mathbf{P} \, d\mathcal{I}
$$

be the entropy four-vector. In correspondence we have the entropy law (7) with the entropy production

$$
\Sigma = -k_B \, c \int_{\Re^3} \int_0^{+\infty} Q \ln f \phi(\mathcal{I}) \, d\mathbf{P} \, d\mathcal{I}.
$$

Neglecting third and higher order terms in non-equilibrium variables we have the H-theorem

$$
\Sigma \geq 0.
$$

Proof - We have in the present case

$$
\Sigma = -\frac{k_B}{c\tau} U_\alpha \int_{\Re^3} \int_0^{+\infty} \left(f_E - f - f_E \, p^\mu q_\mu \frac{1 + \frac{\mathcal{I}}{mc^2}}{bmc^2} \right) \ln f \, p^\alpha \phi(\mathcal{I}) d\mathbf{P} \, d\mathcal{I}. \tag{33}
$$

Let us consider the right hand side of this relation, but with $\ln f_E$ instead of $\ln f$, that is,

$$
-\frac{k_B}{c\tau} U_\alpha \int_{\Re^3} \int_0^{+\infty} \left(f_E - f - f_E \, p^\mu q_\mu \frac{1 + \frac{\mathcal{I}}{mc^2}}{bmc^2} \right) \ln f_E \, p^\alpha \phi(\mathcal{I}) d\mathbf{P} \, d\mathcal{I}. \tag{34}
$$

Thanks to eq. (29) it becomes

$$
\begin{aligned}
-\frac{k_B}{c\tau} U_\alpha \int_{\Re^3} \int_0^{+\infty} & \left(f_E - f - f_E \, p^\mu q_\mu \frac{1 + \frac{\mathcal{I}}{mc^2}}{bmc^2} \right) \cdot \\
& \cdot \left[-1 + \frac{m g_r}{k_B T} - \frac{1 + \frac{\mathcal{I}}{mc^2}}{k_B T} U_\beta p^\beta \right] p^\alpha d\mathbf{P} \, ,
\end{aligned}
$$

which is zero, thanks to (32). Consequently, we can subtract (34) from (33) so that it becomes

$$
\Sigma = -\frac{k_B}{c\tau} U_\alpha \int_{\Re^3} \int_0^{+\infty} \left(f_E - f - f_E \, p^\mu q_\mu \frac{1 + \frac{\mathcal{I}}{mc^2}}{bmc^2} \right) \ln \frac{f}{f_E} \, p^\alpha \phi(\mathcal{I}) d\mathbf{P} \, d\mathcal{I}.
$$

Now we sum to this expression, $(32)_2$ contracted by $-\frac{k_B}{bmc^2}q_\beta$; so it becomes

$$\Sigma = -\frac{k_B}{c\tau}U_\alpha \int_{\Re^3}\int_0^{+\infty}\left(f_E - f - f_E\, p^\mu q_\mu \frac{1+\frac{\mathcal{I}}{mc^2}}{bmc^2}\right)\cdot$$
$$\cdot\left(\ln\frac{f}{f_E} + p^\beta q_\beta \frac{1+\frac{\mathcal{I}}{mc^2}}{bmc^2}\right) p^\alpha \phi(\mathcal{I}) d\mathbf{P}\, d\mathcal{I}.$$

Now, Taylor's expansion of $\ln\frac{f}{f_E}$ around $f = f_E$, dropped at the first order, is given by (26), so that we have

$$\Sigma = \frac{k_B}{c\tau}U_\alpha \int_{\Re^3}\int_0^{+\infty}\frac{1}{f_E}\left(-f_E + f + f_E\, p^\mu q_\mu \frac{1+\frac{\mathcal{I}}{mc^2}}{bmc^2}\right)^2 p^\alpha \phi(\mathcal{I}) d\mathbf{P}\, d\mathcal{I} + O(3)$$

from which it follows $\Sigma \geq 0$.
The classical limit of the present results can be obtained in the same way described for monatomic gases and, also in this case, we find that our BGK relativistic variant converges to the classical BGK.

We observe that the Marle and the Anderson and Witting formulation can be used also in the polyatomic case without any change from the monatomic gas. For this other reason our model seems more physical. In fact the collisional term proposed in this paper changes from (22) to (30) in the passage from monatomic to polyatomic gas. This is what we expected from physical reasons as the collisional terms in polyatomic gas must depend on the internal state, i.e. must depend on $\mathcal{I}$.

The tendency to the equilibrium of the distribution function for our model and comparison between the solutions with the ones using by Marle and Anderson and Witting BGK models will be the subject of a future paper.

Acknowledgments

This paper is dedicated with affection and deep esteem to our friend and colleague Masaru Sugiyama who gave fundamental work in the new approach of Rational Extended Thermodynamics to include the polyatomic and dense gases in the theory. The paper is supported by GNFM of INdAM.

References

[1] Bhatnagar P L , Gross E P and Krook M 1954 *Phys. Rev.* **94** 511
[2] Kremer G M and Cercignani C 2002 *The Relativistic Boltzmann Equation: Theory and Applications* (Basel: Springer)
[3] Müller I and Ruggeri T 1998 *Rational Extended Thermodynamics* 2nd ed. (New York: Springer)
[4] Ruggeri T and Sugiyama M 2015 *Rational Extended Thermodynamics beyond the Monatomic Gas* (Cham, Heidelberg, New York, Dordrecht, London: Springer)
[5] Marle C 1965 *C. R. Acad. Sci. Paris* **260** 6539
[6] Anderson J L and Witting H R 1974 *Physica* **74** 466
[7] Liu I-S, Müller I and Ruggeri T 1986 *Ann. Phys.* **169** 191
[8] Pennisi S and Ruggeri T 2017 *Ann. Phys.* **377** 414
[9] De Groot S R, van Leeuven W A and van Weert Ch G 1980 *Relativistic Kinetic Theory. Principles and Applications* (Amsterdam, New York, Oxford: North-Holland Pub. Co.)
[10] Rezzolla L and Zanotti O 2013 *Relativistic Hydrodynamics* (Oxford: Oxford University Press)
[11] Bellouquid A, Calvo J, Nieto J and Soler J 2012 *J. Stat. Phys.* **149** 284
[12] Takamotu M and Inutsuka S 2010 *Physica A* **389** 4580
[13] Borgnakke C and Larsen P S 1975 *J. Comput. Phys.* **18** 405
[14] Bourgat J F, Desvillettes L, Le Tallec P and Perthame B 1994 *Eur. J. Mech. B-Fluid* **13** 237

IOP Conf. Series: Journal of Physics: Conf. Series **1035** (2018) 012005 doi:10.1088/1742-6596/1035/1/012005

[15] Pavić M, Ruggeri T and Simić S 2013 *Physica A* **392** 1302
[16] Arima T, Mentrelli A and Ruggeri T 2014 *Ann. Phys.* **345** 111
[17] Arima T, Ruggeri T, Sugiyama M and Taniguchi S 2016 *Ann. Phys.* **372** 83

IOP Conf. Series: Journal of Physics: Conf. Series **1035** (2018) 012006 doi:10.1088/1742-6596/1035/1/012006

Effect of crowding on protein-protein association in diffusion-limited regime

Jing Qing, Anpu Chen and Nanrong Zhao
College of Chemistry, Sichuan University, Chengdu 610064, China
E-mail: zhaonanr@scu.edu.cn

Abstract.
In the present paper, we establish a theoretical formalism to study the protein-protein association rate in the framework of diffusion-limited theory. To account for the crowding effect due to the presence of polymer, we particularly take into account the deviation of rotational diffusion of proteins from the Stokes-Einstein-Debye relation. Based on fluid mechanics and depletion theory, a new scaling relation for the retardation factor of the rotational diffusion was proposed. Besides, the crowding-induced interaction energy between the proteins has been also properly introduced into the association rate theory. We apply our theory to calculate the association rate constant of proteins β-lactamase and β-lactamase inhibitor protein in poly(ethylene glycol) solutions. Particular attention is paid to the dependence of the association rate constant on the polymer concentration and polymer molecular weight. The deviation from the simple relation based on Stokes-Einstein approximation is well addressed. We find that our theoretical results show good agreements with the experimental data in the whole concentration region, which demonstrates the validity of our theory.

1. Introduction

Protein association is an important center for various cellular processes such as enzyme catalysis, regulation of immune response by cytokines as well as gene regulation, etc [1, 2]. Understanding the physical principles governing association mechanisms and rate constants and furthermore, realistically modeling them, are crucial to study the relationship between the cellular structure and function. To facilitate experimental studies in vitro, synthetic polymers such as polyethylene glycol (PEG), ficoll, dextran, and poly(vinyl alcohol), are commonly used as a means to mimic molecular crowding in the cell [3, 4, 5]. For instance, crowding effects on protein folding, binding, oligomerization, and protein-protein association have been widely investigated by using polymer solutions [6, 7, 5]. In spite of lots of studies so far, a comprehensive understanding of the crowding effect on protein association is still an open and challenging problem.

To better understand the kinetics of association of two proteins (A and B), it is useful to consider the process as a multistep process [8, 9, 10, 11, 12] below (Fig. 1). The initial step is a collision between two proteins in a solution by virtue of translational diffusion. In the next step rotational diffusion will enable the proteins in the collision complex to find the correct orientation to form the transient complex ($A * B$). At last, the transient complex may evolve to form the final native complex (C) through the transition state, involving desolvation of the interface and exact structural rearrangement. Correspondingly, the overall rate constant of

IOP Conf. Series: Journal of Physics: Conf. Series **1035** (2018) 012006 doi:10.1088/1742-6596/1035/1/012006

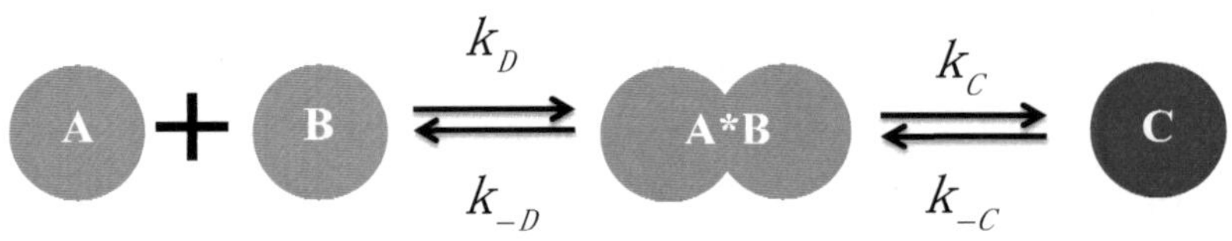

Figure 1. A schematic for the association process. Two proteins are modeled by two globular particles A and B, diffusing to form transient complex $A*B$ and then leading to native complex C by conformer rearrangement.

association is given by

$$k_{\text{on}} = \frac{k_D k_C}{k_{-D} + k_C}, \tag{1}$$

where k_D and k_{-D} represent the rate constants of transient complex formation and dissociation and k_C is the rate constant at which the transient complex evolves to form the final complex. Both the diffusion step and the conformational step can be rate-limiting of protein association. Basically, there is no simple way to determine whether a given protein complex is controlled by diffusion, conformational changes or mixed range. Nevertheless, many experimental [13] observations reveal that in most cases, the conformational rearrangement is fast relative to the dissociation of transient complex (i.e., $k_C \gg k_{-D}$), thus leading to a diffusion controlled regime $k_{\text{on}} \simeq k_D$.

So far, great attention has been paid to investigating diffusion-controlled association rate. For simplicity, by neglecting the rotational step, Smoluchowski equation [14] relates k_D and the translational diffusion coefficients according to

$$k_D = 4\pi \bar{r}_h \bar{D}_t, \tag{2}$$

where $\bar{r}_h$ and $\bar{D}_t$ represent the sum of the hydrodynamic radius r_h and translational diffusion coefficient D_t of proteins A and B. In simple liquids or dilute agents, D_t can be estimated based on Stokes-Einstein (SE) relation i.e., $D_t = k_B T/6\pi\eta_{\text{macro}} r_h$ where η_{macro} denotes the macroscopic viscosity of the solution, k_B is the Boltzmann constant and T is the temperature. Further assuming two proteins undergoing association having similar size, it is easy to derive the relative association rate constant obeys the following simple linear relation:

$$\frac{k^0_{\text{on}}}{k_{\text{on}}} = \frac{\eta_{\text{macro}}}{\eta_0}, \tag{3}$$

where "0" denotes the value in pure solvent.

Recently, many experiments [15, 16] have been applied to investigate the protein-protein association in polymer solutions by using stopped-flow fluorescence spectrometer. Interestingly, experimental studies have revealed that the association rate in polymer solutions will seriously deviate from the simple relation based on SE approximation as given by Eq. (3). The deviation is almost negative; i.e., the decrease in association rate with increased viscosity is less than predicted, indicating a faster rate constants in crowding and viscous polymer solutions. Furthermore, the negative deviation increases with increasing polymer concentration and molecular weight. The understanding of such a novel phenomenon can be very important for modeling processes in vivo. To the best of our knowledge, however, such an understanding is still lacking.

Motivated by the above consideration, in the present paper, we will establish a theoretical formalism to analyze the association rate constant for proteins in polymer solutions. The crowding effect due to the presence of polymer will be properly introduced. Based on the schematic of the theory of diffusion-limited association (DLA), both translational and rotational steps are included. The translational diffusion coefficient is assumed to satisfy the SE relation, while the deviation of rotational diffusion coefficient from the Stokes-Einstein-Debye (SED) relation will be seriously taken into account. Adopting the fluid mechanics and depletion theory we proposed a new scaling law for the rotational retardation factor. Furthermore, we introduce the crowding-induced interaction energy between proteins which can be specified based on the thermodynamic analysis for the change of free energy for both reactants and transient complex from dilute to crowding solutions. Finally, we reach the formulation for the association rate constant, in explicit relationship to the property of polymer solutions and the hydrodynamic radii of the proteins. We apply our theory to study the association of proteins β-lactamase (TEM) and β-lactamase inhibitor protein (BLIP) in PEG solutions. The relative association rate constant in PEG solutions with different PEG molecular weight and varying concentration will be explicitly evaluated. Through evaluating the dependence of the association rate on the macroscopic viscosity, we quantitatively analyze the negative deviation from SE estimation. We compare our theoretical results with the experimental data, and good agreements have been achieved.

The paper is organized as follows. Firstly, we introduce the DLA theory for evaluating the protein-protein association rate constant. The crowding effects due to the deviation of rotational diffusion coefficient from SED relation as well as the crowding-induced interaction energy between proteins are properly taken into account. Secondly, we apply the theory to study the association of TEM and BLIP in PEG solutions. The theoretical results will be compared with the experimental data. At last, we conclude the paper.

2. Model and theory

According to the experimental observations, the protein-protein association rate constant in most cases occur in the diffusion-controlled regime. Therefore, it is reasonable for us to turn to DLA theory. The theory starts from the combined rotation-translation diffusion equation for a pair of particles (spheres) moving in a solution. To account for the nonuniform reactivity of the proteins and the fact that reaction should occur at the proper mutual orientation, the analytic expression for the fully DLA reaction rate constant has been well established based on Szabo approach [17], which yields [18]

$$k_D = 4\pi \left(D_{t,A} + D_{t,B}\right)\left(r_h^A + r_h^B\right)\left[F_A \zeta_B \tan\left(\alpha_B/2\right) + F_B \zeta_A \tan\left(\alpha_A/2\right)\right], \tag{4}$$

where $\zeta_i = \sqrt{[1 + D_{r,i}(r_h^A + r_h^B)^2/(D_{t,A} + D_{t,B})]/2}$ where D_r denotes the rotational diffusion coefficient, and $F_i = \sin^2(\alpha_i/2)$ with α_i being the symmetric reactive region. According to the above expression, k_D is determined by both translational and rotational diffusion. Note that recent experimental studies for diffusion of molecular probes in complex polymer solutions, reveal a very novel *decoupling* phenomenon between the translational and rotational diffusion [19, 20]. Namely, it was found experimentally that, for diffusion of proteins in polymer solutions, the SE relation for the translational diffusion remains well satisfied, while rotational diffusion exhibits large deviations from the traditional SED relation given by $D_r = k_B T/8\pi\eta_{\text{macro}} r_h^3$. Therefore, it seems reasonable for us to introduce SE relation for D_t and the proper modification for D_r into Eq. (4). The modification concerning the rotational diffusion coefficient will be provided later. Furthermore, if we assume $r_h^A \simeq r_h^B$ and also $\alpha_A \simeq \alpha_B$, we can derive the relative rate constant (i.e., k_D^0/k_D with k_D^0 denotes the value in pure solvent), given by

$$\frac{k_D^0}{k_D} = \frac{\eta_{\text{macro}}}{\eta_0}\sqrt{5\Big/\left(2 + \frac{3}{R^{\text{rot}}}\frac{\eta_{\text{macro}}}{\eta_0}\right)}, \tag{5}$$

IOP Conf. Series: Journal of Physics: Conf. Series **1035** (2018) 012006 doi:10.1088/1742-6596/1035/1/012006

where $R^{\rm rot} \equiv D_r^0/D_r$ represents the retardation factor of rotational diffusion, i.e., the ratio of rotational diffusion coefficient in pure water to that in polymer solutions. Obviously, the relative rate constant given above is expressed explicitly in terms of both macroscopic theory $\eta_{\rm macro}$ and the rotational retardation factor $R^{\rm rot}$. As a limitation case, supposing that SED relation is also valid, such that $R^{\rm rot} = \frac{\eta_{\rm macro}}{\eta_0}$. Then, we have immediately Eq. (3) which is just the association rate constant under SE approximation.

The macroscopic viscosity for polymer solution is certainly determined by the structure of the polymer solutions, including the polymer size, concentration, and even the statistical length scale, such as mesh length ξ. Here we adopt the promising empirical scaling relation for $\eta_{\rm macro}$ proposed by Holyst et al. [21],

$$\frac{\eta_{\rm macro}}{\eta_0} = \exp\left[\frac{\gamma}{RT}\left(\frac{R_h}{\xi}\right)^{\beta}\right], \tag{6}$$

where R denotes the gas constant, γ is a phenomenological parameter varying from system to system, β is also an empirical parameter, equal to 0.78 for most polymer solutions [22], R_h is the hydrodynamic radius of the polymer molecule, and ξ is the correlation length depending on polymer concentration ϕ, following

$$\xi = R_g\left(\frac{\phi}{\phi^*}\right)^{-3/4}, \tag{7}$$

where R_g is the polymer gyration radius and ϕ^* is the overlap value for crossover from the dilute to semidilute regime, determined by the property of the polymer, according to

$$\phi^* = \frac{3M_w}{4\pi d N_A R_g^3} \tag{8}$$

with M_w being the polymer molecular weight, d the polymer mass density, and N_A the Avogadro constant.

The rotational retardation factor $R^{\rm rot}$ for protein in polymer solutions, which can heavily deviate from SED relation as mentioned above, which therefore will lead to the relative association rate constant as expressed in Eq. (5) rather nontrivial. Based on fluid mechanics [23], $R^{\rm rot}$ can be evaluated explicitly. When a sphere is rotating in a quiescent fluid, the low Reynolds number flow is solely driven by the shear stress. The fluid velocity field around the particle is assumed to be incompressible and obey the Navier-Stokes equations but now with space-dependent viscosity associated to the depletion effect. Via solving the field equations with proper boundary conditions, we can get the shear stress of the fluid at the probe surface. This allows us to calculate the shear-induced torque Γ exerted on the probe by polymer solutions. Assuming the probe is rotating with angular velocity Ω, the rotational diffusion coefficient thus follows [24] $D_r \equiv k_B T\Omega/\Gamma$. As demonstrated in Ref. [25, 26], the rotational diffusion coefficient deviated from the SED relation remarkably. The retardation factor $R^{\rm rot}$ unexpectedly is not sensitive to the changing of the relative macroscopic viscosity $\eta_{\rm macro}/\eta_0$, while is relevant to the thickness of the so-called depletion layer δ. Moreover, $R^{\rm rot}$ strikingly exhibits an apparent scaling behavior with respect to the ratio between the probe hydrodynamic radius and the depletion thickness, i.e., r_h/δ, as shown in Fig. 2. Readers may refer to Ref. [27] for more details. Here, for simplicity, we only quote the final results for the scaling relation of the rotational retardation factor as follows:

$$R^{\rm rot} = \frac{D_r^0}{D_r} = \exp\left[a\left(r_h/\delta\right)^b\right], \tag{9}$$

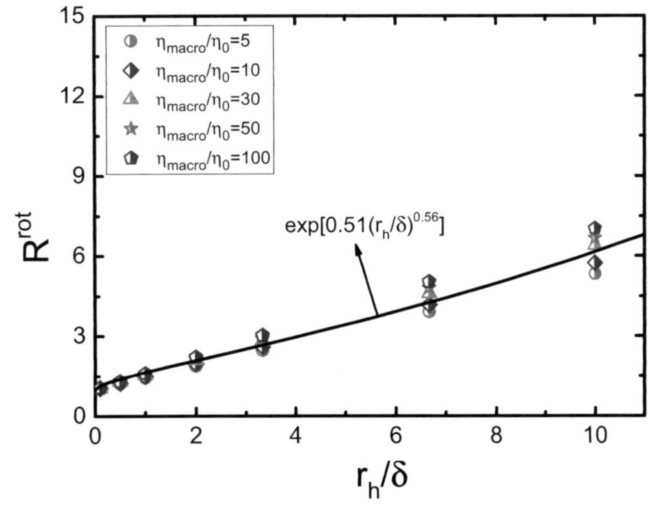

Figure 2. Rotational retardation factor $R^{\rm rot}$ as a function of the length scale r_h/δ, under different macroscopic viscosity $\eta_{\rm macro}$. The solid line represents the scaling relation as expressed in the figure.

where a and b are fitting parameters given by $a \simeq 0.51$ and $b \simeq 0.56$. We emphasize this scaling law is a general result which is obtained without going into details of any specific information of molecular probes and polymer solutions. The values of the parameters a and b seem robust and universal. Comparing with the retardation factor for translational diffusion, which is determined by the effective hydrodynamics $R_{\rm eff}$ divided by polymer correlation length ξ, i.e., $R_{\rm eff}/\xi$, $R^{\rm rot}$ for rotational diffusion, however, is determined by a very different length scale r_h/δ. Note that this scaling can recover the limitations of non-retardation $R^{\rm rot} \to 1$ as $\delta/r_h \to \infty$, and full retardation $R^{\rm rot} \to \infty$ as $\delta/r_h \to 0$. In our opinion, the generality of the scaling relation (9) is basically guaranteed in a certain degree.

The depletion thickness δ in the formulation (9) is a polymer solution parameter which has a meaning of the depletion thickness near a flat wall in semidilute solutions. As it is well-known, according to many recent experimental and theoretical studies, for colloidal particles suspended in polymer solutions, there appears a depletion layer in the vicinity of the particle surface, introducing a non-homogenous concentration distribution [28, 29]. The depletion thickness characterizes in what extent the non-homogeneity of the polymer concentration sustains until it finally reaches the bulk value. According to the mean-filed approximation [30], δ can be given by

$$\frac{1}{\delta^2} = \frac{\pi}{4R_g^2} + \frac{1}{\xi_{\rm FH}^2}, \tag{10}$$

where $\xi_{\rm FH}$ is the Flory-Huggins correlation length. In a good solvent situation, $\xi_{\rm FH}$ takes the specific form of $\xi_{\rm FH}^{-2} = -3\,[\ln(1-\phi)]\,\sigma^{-2}$ with σ being the statistical segment length of polymer chains (for example, $\sigma \simeq 0.7$ nm for PEG). Clearly, in the dilute limit, $\xi_{\rm FH} \to \infty$, such that the depletion thickness $\delta \to 2R_g/\sqrt{\pi}$. While in the semidilute regime, δ depends on both polymer gyration radius R_g and polymer concentration ϕ. Obviously, δ is a function of both polymer concentration ϕ and polymer size. Fig. 3 presents dependence of depletion thickness on ϕ and R_g.

IOP Conf. Series: Journal of Physics: Conf. Series **1035** (2018) 012006 doi:10.1088/1742-6596/1035/1/012006

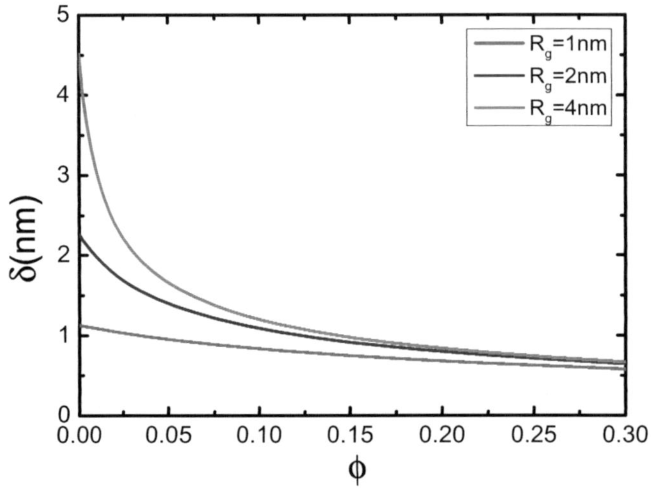

Figure 3. The depletion thickness δ as a function of volume fraction ϕ under different polymer gyration radius R_g.

Furthermore, taking into account that in the presence of polymer solutions, the association process should be subject to the crowding effect, quantified by crowding-induced interaction energy $\Delta\Delta G_c^*$ between proteins. Thus, the association rate constant in crowding environment will company such an extra energy contribution based on the original diffusion-limited formulation (5). As a result, we can express the relative association rate constant as

$$\frac{k_{\rm on}^0}{k_{\rm on}} = \frac{k_D^0}{k_D} \exp\left[\frac{\Delta\Delta G_c^*}{RT}\right] = \frac{\eta_{\rm macro}}{\eta_0}\sqrt{5/\left(2+\frac{3}{R^{\rm rot}}\frac{\eta_{\rm macro}}{\eta_0}\right)} \exp\left[\frac{\Delta\Delta G_c^*}{RT}\right]. \tag{11}$$

Obviously, the value of crowding-induced interaction energy might depends on the path of the reaction, the properties of the reactants as well as the polymer solutions. Owing to Minton's work [31], $\Delta\Delta G_c^*$ can be derived based on a reaction cycle as shown in Fig. 4. Briefly, $\Delta\Delta G_c^*$ has the meaning of the difference in transfer free energy between the transient complex and the two separated proteins, from a dilute solution to a crowded solution, i.e.,

$$\Delta\Delta G_c^* = \Delta G^c - \Delta G^0 = \Delta G_{A*B}^{dil\to crd} - \left(\Delta G_A^{dil\to crd} + \Delta G_B^{dil\to crd}\right), \tag{12}$$

where ΔG^0 denotes the free energy for association of A and B in the absence of crowders and ΔG^c denotes the free energy for that in the presence of crowders. $\Delta G_i^{dil\to crd}$ represents the change of free energy for molecular probes from dilute to crowded solutions, which is determined by the relative size of molecular probes and crowders and concentration of crowders. According to the scaled particle theory (SPT) of hard particle fluids proposed by Lebowitz and co-workers [32, 33], we have the following analytical expression for $\Delta G_i^{dil\to crd}/RT$:

$$\frac{\Delta G_i^{dil\to crd}}{RT} = -\ln(1-\phi)+\left(\frac{\phi}{1-\phi}\right)\left(z_i^3+3z_i^2+z_i\right)+\left(\frac{\phi}{1-\phi}\right)^2\left(3z_i^3+\frac{9}{2}z_i^2\right)+\left(\frac{\phi}{1-\phi}\right)^3\left(3z_i^3\right), \tag{13}$$

where $z_i = r_h^i/R_g$. The expression above provides the explicit relationship between the change of free energy $\Delta G_i^{dil\to crd}$ with respect to the polymer concentration, as well as the hydrodynamic radii of both reactant A, B and transient complex $A * B$.

IOP Conf. Series: Journal of Physics: Conf. Series **1035** (2018) 012006 doi:10.1088/1742-6596/1035/1/012006

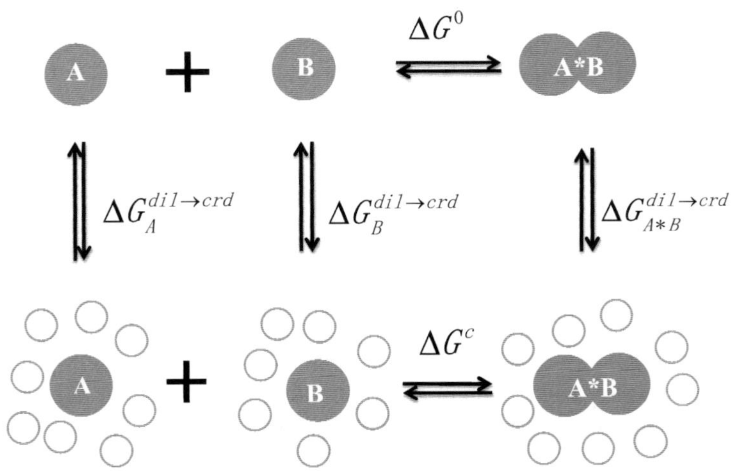

Figure 4. An illustration of the thermodynamic cycle: protein-protein association in dilute and crowded solutions.

Table 1. The basic parameters for PEG solutions.

PEG	R_g (nm)	R_h (nm)	ϕ^*
3.35K	2.44	1.49	0.081
6K	3.43	2.08	0.052
8K	4.05	2.45	0.042

Eq. (11) is our main equation. Inserting the macroscopic viscosity η_{macro} Eq. (6), rotational retardation factor R^{rot} Eq. (9) as well as the crowding-induced interaction energy $\Delta\Delta G_c^*$ Eq. (12), we can explicitly analyze the protein-protein association rate constant in polymer solutions.

3. Application to PEG systems

We apply the theory to quantitatively evaluate the association rate constants of globular proteins, TEM and BLIP, in PEG solutions. The hydrodynamic radii of TEM and BLIP are roughly equal to be $r_h \simeq 2.2$ nm. For PEG systems, the gyration and hydrodynamic radii of the coils, i.e., R_g and R_h can be estimated according to the molecular weight M_w via [34, 35] $R_g = 0.0215M_w^{0.583}$ and $R_h = 0.0145M_w^{0.571}$. γ is estimated by fitting experimental data, which has been reported to be 4.0 kJ/mol [22]. In addition, the mass density d=1126 kg/m^3 and the segmental statistical length σ=0.7 nm [36]. In order to compare our theory with experimental work, we correspondingly consider the association in PEG solutions with several molecular weights M_w=3.35K (3.35 kg/mol), 6K (6 kg/mol) and 8K (8 kg/mol) respectively. The polymer concentration varies from 0 to 0.3 or so. For PEG solutions, the basic parameters are listed in Table 1.

In order to specify the theoretical formulation of the association rate constant k_{on} based

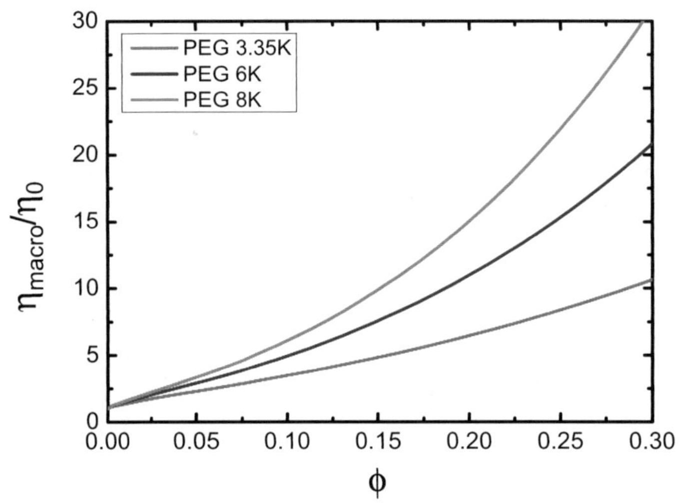

Figure 5. Relative macroscopic viscosity $\eta_{\text{macro}}/\eta_0$ plotted as a function of polymer concentration ϕ.

on Eq. (11), we have provided all the necessary parameters except the hydrodynamic radius of the transient complex, i.e., r_h^{A*B}. Note that r_h^{A*B} is just a model parameter coming from the thermodynamic cycle which is used to determine the change of free energy of the transient complex from dilute to crowded solutions, i.e., $\Delta G_{A*B}^{dil\to crd}$, according to Eq. (13). Evidently, r_h^{A*B} stands for an effective hydrodynamic radius, and can be determined only by fitting experiments.

Now we are ready to analyze the association rate constant of TEM and BLIP in PEG solutions. Firstly, in Fig. 5 we plot the relative macroscopic viscosity $\eta_{\text{macro}}/\eta_0$ as a function of polymer concentration ϕ according to Eq. (6), for PEG 3.35K, 6K and 8K with ϕ ranging from 0 to 0.3. Obviously η_{macro} increases with polymer concentration for solution of PEG with a certain value of R_g. Also, it increases with R_g at a fixed value of ϕ. Then, it is straightforward for us to evaluate the dependence of the association rate constant k_{on} on the macroscopic viscosity η_{macro} of PEG solutions. By adjusting the single parameter, i.e., the effective hydrodynamic radius of the transient complex r_h^{A*B}, with all other parameters of the system (see Table 1 and the context), we compare our theoretical results with the available experimental data [8]. As a result, we have Fig. 6, where both our theoretical results (solid lines) and the experimental data (symbols) are shown. Herein, the corresponding fitting parameters are estimated as $r_h^{A*B} \simeq$ 2.9nm, 2.5nm and 2.2nm in PEG 3.35K, 6K, and 8K, respectively, which are in similar size as the proteins TEM and BLIP. Evidently, our theory can agree with the experimental observations very well in the whole concentration region, which demonstrates the validity of our theory in a certain degree.

As shown in Fig. 6, the association rate constant slightly increases with the macroscopic viscosity, and then reaches a plateau at high concentration region. Eq. (3) is also plotted (dashed line). Evidently, the association rate constant deviates from the simple relation based on SE approximation (Eq. (3)), and the deviation is definitely negative indicating the association rate is much faster than the estimation according to SE estimation. Furthermore, the deviation becomes more and more remarkable along with the increase of the polymer concentration (or macroscopic viscosity η_{macro}). This phenomena can be well understood by using our model based on Eq. (11), where two important terms, namely rotational retardation factor R^{rot} and

IOP Conf. Series: Journal of Physics: Conf. Series **1035** (2018) 012006 doi:10.1088/1742-6596/1035/1/012006

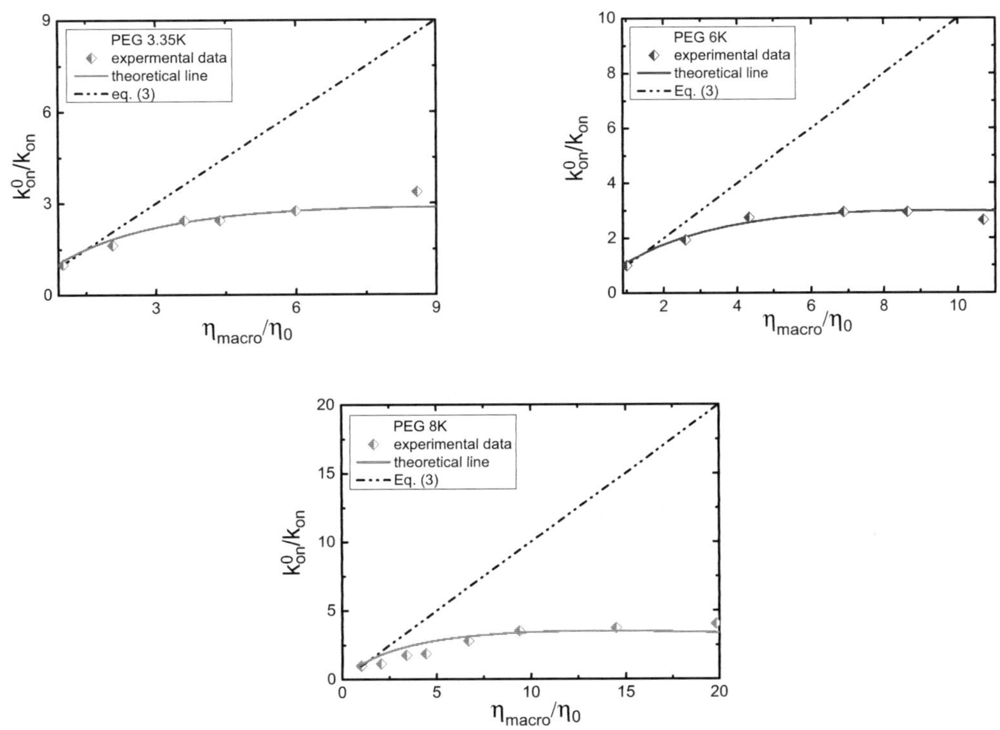

Figure 6. Relative association rate constant $k^0_{\mathrm{on}}/k_{\mathrm{on}}$ as a function of relative macroscopic viscosity of solutions of PEG with different molecular weights.

the crowding-induced interaction energy $\Delta\Delta G^*_c$ between proteins are properly involved. Firstly, as mentioned above, R^{rot} deviates from estimation of SED relation, leading to a much faster rotational diffusion in PEG solutions due to the depletion effect. Secondly, the presence of $\Delta\Delta G^*_c$ can be another factor which will inevitably enhance the association rate. Therefore, Eq. (11) captures the main factors which characterize the association rate in PEG solutions.

4. Conclusions

In the present work, we proposed a theoretical framework to understand the crowding effect on protein-protein association rate constant k_{on}. Based on the rotation-translation DLA theory Eq. (4), the diffusion-controlled association rate k_D can be determined. We assume the translational diffusion coefficient obeys SE relation, while the rotational diffusion deviates remarkably from SED relation in polymer solutions. We introduce the rotational retardation factor R^{rot} and then obtain Eq. (5). R^{rot} follows a new scaling form as demonstrated by Eq. (9) and Fig. 2, which is derived from the fluid mechanics. Furthermore, we consider the modification associated to the crowding-induced interaction energy $\Delta\Delta G^*_c$ between proteins, which can be made specific based on the thermodynamic analysis for the change of free energy $\Delta G_i^{dil\to crd}$ from a dilute solution to a crowded solution for proteins and transient complex, as given by Eq. (13). Finally, we have Eq. (11) which can be used as the basic formulation to evaluate the rate constant k_{on} for protein-protein association in polymer solutions.

We apply our theory to analyze the TEM-BLIP association in PEG solutions by keeping the hydrodynamic radius r_h^{A*B} of the transient complex as a fitting parameter. The relative association rate constant under different molecular weights of PEG, and its dependence on polymer concentration or macroscopic viscosity, have been extensively analyzed. As a result, our theoretical results can well reproduce the experimental data in the whole concentration region with very reason fitting parameter r_h^{A*B}, which demonstrates the validity of our theory. We would like to emphasize our theory might capture the main factors which lead to the large negative deviation from the simple relation of $k_{\rm on}^0/k_{\rm on} = \eta_{\rm macro}/\eta_0$. We find that the negative deviation of rotational diffusion coefficient from SED relation due to the depletion effect, as well as the crowding-induced interaction energy, can account for the peculiarly fast protein-protein association rate in the polymer solutions. Since protein association in complex crowding environment is of ubiquitous importance in many cellular processes, we hope our work might be helpful to gain more perspectives in relevant research fields.

Acknowledgements
This work was supported by National Natural Science Foundation of China (Grant No. 21673145).

References

[1] Mourão M A, Hakim J B and Schnell S 2014 *Biophys. J.* **107** 2761
[2] Tabaka M, Kalwarczyk T and Holyst R 2014 *Nucleic Acids Res.* **42** 727
[3] Lavalette D, Tetreau C, Tourbez M and Blouquit Y 1999 *Biophys. J.* **76** 2744
[4] Tabaka M, Kalwarczyk T, Szymanski J, Hou S and Holyst R 2014 *Frontiers in Physics* **2** 54
[5] Zhou, H X 2013 *FEBS. Lett.* **587** 1053
[6] Dupuis N F, Holmstrom E D and Nesbitt D J 2014 *Proc. Natl. Acad. Sci.* **111** 8464
[7] Phillip Y, Kiss V and Schreiber G 2012 *Proc. Natl. Acad. Sci.* **109** 1461
[8] Kozer N, Kuttner Y Y, Haran G and Schreiber G 2007 *Biophys. J.* **92** 2139
[9] Schreiber G, Haran G and Zhou H X 2009 *Chem. Rev.* **109** 839
[10] Zhou H X 2010 *Q. Rev. Biophys.* **43** 219
[11] Zhou H X and Bates P A 2013 *Curr. Opin. Struc. Biol.* **23** 887
[12] Zhou H X and Pang X 2018 *Chem. Rev.* **118** 1691
[13] Qin S, Pang X and Zhou H X 2011 *Structure* **19** 1744
[14] Smoluchowski M V Z 1918 *Phys. Chem.* **9** 129
[15] Kuttner Y Y, Kozer N, Segal E, Schreiber G and Haran G 2012 *J. Am. Chem. Soc.* **127** 15138
[16] Kozer N and Schreiber G 2004 *J. Mol. Biol.* **336** 763
[17] Shoup D, Lipari G and Szabo A 1981 *Biophys. J.* **36** 697
[18] Zhou H X 1993 *Biophys. J.* **64** 1711
[19] Roos M, Ott M, Hofmann M, Link S, Rossler E, Balbach J, Krushelnitsky A and Saalwachter K 2016 *J. Am. Chem. Soc.* **138** 10365
[20] Edmond K V, Elsesser M T, Hunter G L, Pine D J and Weeks E R 2012 *Proc. Natl. Acad. Sci. U. S. A.* **109** 17891
[21] Wisniewska A, Sozanski K, Kalwarczyk T, Kedra-Krolik K, Pieper C, Wieczorek S A, Jakiela S, Enderlein J and Holyst R 2014 *Polymer* **55** 4651
[22] Sozanski K, Wisniewska A, Kalwarczyk T and Holyst R 2013 *Phys. Rev. Lett.* **111** 228301
[23] Fan T-H, Xie B and Tuinier R 2007 *Phys. Rev. E.* **76** 051405
[24] Saffman P G and Delbruck M 1975 *Proc. Nat. Acad. Sci.* **72** 3111
[25] Lavalette D, Hink M A, Tourbez M, Tetreau C and Visser A J 2006 *Eur. Biophys. J* **35** 517
[26] Li C, Wang Y and Pielak G J 2009 *J. Phys. Chem. B.* **113** 13390
[27] Qing J, Chen A and Zhao N 2017 *Phys. Chem. Chem. Phys.* **19** 32687
[28] Ziebacz N, Wieczorek S A, Kalwarczyk T, Fialkowski M and Holyst R 2011 *Soft Matter* **7** 7181
[29] Feng X, Chen A, Wang J, Zhao N and Hou, Z 2016 *J. Phys. Chem. B.* **120** 10114
[30] Fleer G J, Skvortsov A M and Tuinier R 2003 *Macromolecules* **36** 7857
[31] Minton A P 2006 *J. Cell Sci.* **119** 2863
[32] Reiss H, Frisch H L and Lebowitz J L 1959 *J. Chem. Phys.* **31** 369
[33] Lebowitz J L, Helfand E and Praestgaard E 1965 *J. Chem Phys* **43** 774
[34] Devanand K and Selser J C 1991 *Macromolecules* **24** 5943
[35] Linegar K L, Adeniran A E, Kostko A F and Anisimov M A 2010 *Colloid. J.* **72** 279
[36] Dong Y, Feng X, Zhao N and Hou Z 2015 *J. Chem. Phys.* **143** 024903

IOP Conf. Series: Journal of Physics: Conf. Series **1035** (2018) 012007 doi:10.1088/1742-6596/1035/1/012007

Numerical simulation of ionospheric convection with a global MHD simulation

Satoko Saita[1], Shigeru Fujita[2], Akira Kadokura[3], Takashi Tanaka[4] and Yoshimasa Tanaka[3]

[1] National Institute of Technology, Kitakyushu College, Kitakyushu, Japan
[2] Meteorological College, Japan Meteorological Agency, Chiba, Japan
[3] National Institute of Polar Research, Tokyo, Japan
[4] International Center for Space Weather Science and Education, Kyushu University, Fukuoka, Japan

E-mail: saita@kct.ac.jp

Abstract. A case study was conducted for simulated processes of magnetosphere-ionosphere (M-I) coupling process using a magnetohydrodynamic (MHD) simulation code developed by Tanaka (2010). We combined the MHD simulation and solar wind parameters derived from the ACE satellite, and compared the ionospheric $\mathbf{E} \times \mathbf{B}$ plasma drift obtained from the global MHD simulation and that obtained from the SuperDARN HF Radar Network. The simulated plasma drift was not always reproducible under a southward interplanetary magnetic field (IMF) condition. We believe that the M-I boundary conditions in the global MHD simulation includes insufficient factors for the M-I coupling process.

1. Introduction

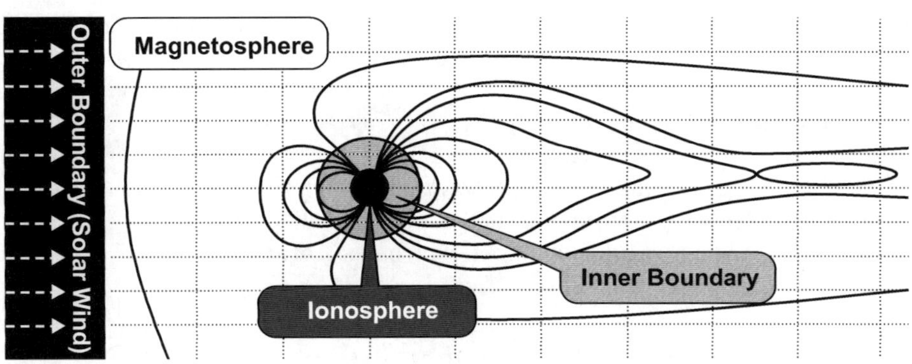

Figure 1. A schematic diagram of the meridional magnetosphere noon-midnight section with the outer boundary (the solar wind) and the inner boundary (the ionosphere).

The Earths magnetic field with north and south poles prevents the solar wind from directly hitting the Earth. The region where the Earths magnetic field deflecting the solar wind is

IOP Conf. Series: Journal of Physics: Conf. Series **1035** (2018) 012007 doi:10.1088/1742-6596/1035/1/012007

called "the magnetosphere". The night-side magnetosphere is dragged out while the sunward side magnetosphere is compressed. Figure 1 shows a schematic diagram of the magnetosphere blasted by the solar wind. In this study, we use a magnetohydrodynamic (MHD) simulation of the magnetosphere-ionosphere coupling system, which enables us to examine the magnetospheric response to the solar wind.

MHD simulations have been developed to reproduce the magnetosphere from the early 1980's [1,2]. An increase in performance of computers in recent years allowed MHD calculations to employ minimum meshes and to reproduce magnetospheric phenomena [3,4]. Tanaka (2003) developed a high-resolution global MHD simulation that employs a finite volume (FV) total variation diminishing (TVD) scheme with an unstructured triangular grid system [5].

The global MHD model has been improved on the response of the Earth's magnetosphere to variations in solar wind [6]. Thus, today, this model consists the magnetosphere-ionosphere coupling model and fetches solar wind parameters as an outer boundary condition of the magnetosphere model. As shown in Figure 1, the model sets one outer boundary at a distance of 20 times the radius of the Earth (20 R_E; R_E denotes Earth radius) from the Earth's center in the direction toward the sun. We use physical quantities associated with the solar wind as an outer boundary condition for the magnetosphere model. A second outer boundary is set at 200 R_E from the Earth's center in the direction opposite the sun; the boundary condition is configured so that the gradients of all physical quantities vanish.

Meanwhile, the inner boundary of the magnetosphere lies 3 R_E from the center of the Earth. The system of MHD equation is solved in the magnetosphere. Physical quantities from the magnetosphere at the inner boundary are projected to the ionosphere along magnetic field lines. Ohm's law is solved to match the divergence of the Pedersen and Hall currents with the field-aligned current (FAC) in the ionosphere. Physical quantities from the magnetosphere model and those from the ionosphere are interchanged with via some equations describing magnetosphere-ionosphere (M-I) interactions at the inner boundary (equations (1), (2), (3), and (4), as discussed in detail in Section 2). The equations include empirically determined parameters representing effects of the plasma pressure, temperature, and electric currents to the ionospheric electrical conductivity. Therefore, we expect the ionospheric conductivity and the ionospheric electric potential to vary depending on the empirically determined parameters.

To improve the existing model, we have to compare calculated and observed values of the conductivity and the electric potential of the ionosphere to estimate the optimal combinations of the empirically determined parameters. A realistic description of electromagnetic phenomena associated with auroras has become another task for this study.

In the present study, we reproduced the ionospheric potential by running the global MHD simulations while changing the empirically determined parameters. In particular, our discussion focuses on ionospheric electric potential variations in cases where the magnetic field of the solar wind is directed southward, a condition under which *substorms* — sudden releases of energy contained in the magnetosphere from solar wind — readily occur.

2. Magnetosphere-Ionosphere (M-I) Coupling Process

At the inner boundary of the magnetosphere, we compute the magnetic field structure and the electric current distribution of the magnetosphere, and then determine the ionospheric conductivity σ and potential distribution Φ_I from the magnetic-field-aligned current, $J_{\parallel}$. Then, we determine the background magnetic field $\mathbf{B}$ and the momentum m, followed by the potential distribution of the magnetosphere Φ_M from Φ_I.

$$\nabla \cdot \sigma \nabla \Phi_I = G_m(\mathrm{rot}\mathbf{B}_1 \cdot \mathbf{n}_b) = J_{\parallel}, \tag{1}$$

$$\sigma = \sigma_{\mathrm{EUV}} + k_2 \sigma_{\mathrm{Diff}}(P, \rho) + k_3 \sigma_J (J_{\parallel}), \tag{2}$$

$$\Phi_M = \Phi_I - k_4 f_2(J_{\parallel})|J_{\parallel}|, \tag{3}$$

$$\mathbf{m} - (\mathbf{m}\cdot\mathbf{n}_b)\mathbf{n}_b = -\rho\nabla\Phi_M \times \mathbf{B}/B^2, \tag{4}$$

where $\mathbf{B}_1$ is the variable component of magnetic field, G_m is a geometrical factor associated with the mapping along field lines from $r = 3\ R_E$ to $r = 1\ R_E$, and $\mathbf{n}_b$ is an outward unit vector along $\mathbf{B}$. Equation (1) indicates that $J_{\parallel}$ from the ionosphere is a current flowing from the magnetosphere to the ionosphere that arises from excesses or deficiencies in the horizontal current flow in the ionosphere, while $J_{\parallel}$ from the magnetosphere arises from shearing motion $(\mathrm{rot}\mathbf{B}_1 \cdot \mathbf{n}_b)$ of the plasma in response to magnetic field variations.

Equation (2) indicates that several ionospheric conductivity components (σ_{EUV}, σ_{Diff}, σ_J) can increase the ionospheric conductivity σ. σ_{EUV}, σ_{Diff}, σ_{J} mean a component due to extreme ultraviolet rays from the sun, a component due to diffuse auroras determined by the temperature and pressure of the magnetosphere, and a component due to discrete auroras determined by the upward-pointing $J_{\parallel}$.

Equation (3) says that Φ_I, augmented by the Joule dissipation of currents along the field lines $(k_4 f_2(J_{\parallel})|J_{\parallel}|)$, is reflected as Φ_M. Equation 4 is a rearrangement of Ohm's law; in combining equations (1), (2), and (3), we have solved Ohm's law to ensure that the divergence of the current flowing horizontally in the magnetosphere agrees with $J_{\parallel}$.

Coefficients such as k_2, k_3, and k_4 in the equations above are scaling factors. f_2 takes the value 1 for upward-pointing $J_{\parallel}$ and 0 for downward-pointing $J_{\parallel}$. k_2, k_3, and k_4 are the empirically determined parameters representing effects of the plasma pressure, temperature, and electric currents to the ionospheric electrical conductivity.

Here, we focus on the electric potential in the ionosphere; we report the results of tests to assess the sensitivity of the global MHD simulation model to the empirically determined parameters.

3. Event Selection

Solar wind parameters, namely solar wind velocity, proton temperature, proton density and mean magnetic field in the solar wind, were used as an outer boundary condition. The solar magnetic field that spreads into interplanetary space is called Interplanetary Magnetic Field (IMF). The ACE satellite provides in-situ data of the solar wind parameters between Earth and Sun. For this study, we selected the interval that strong magnetospheric and ionospheric convections caused.

Figure 2 shows IMF on 11th September 2012 is directed to the southward and implies the ionospheric plasma convection is strong. Geomagnetic activity is evaluated using the geomagnetic auroral electrojet (AE) index [7], from the World Data Center for Geomagnetism, Kyoto (`http://swdcwww.kugi.kyoto-u.ac.jp/`). In this study, we used the AE index for the proxy of substorm activities and found that AE index on 11th September was low. Then the IMF Bz and the AE index indicate that strong magnetospheric and ionospheric convection caused, however, no auroras occurred in the interval.

Because the ACE satellite orbits at the Lagrangian point L_1, we can expect a time delay between the outer boundary and ACE data. To compare the ionospheric convection obtained from simulation data to actual observed values, we adjusted the time delay from the ACE satellite to the magnetopause.

In this paper, the x-axis points toward the Sun, the z-axis points north, and the y-axis is chosen to satisfy the right-handed system.

4. Simulation Results

The simulation started from a stationary state with solar wind parameters of the proton density, $N = 3$ /cc, the solar wind velocity, $V_{SW} = 300$ km/s, the y-component of IMF, $B_Y = -1.6$ nT,

IOP Conf. Series: Journal of Physics: Conf. Series **1035** (2018) 012007 doi:10.1088/1742-6596/1035/1/012007

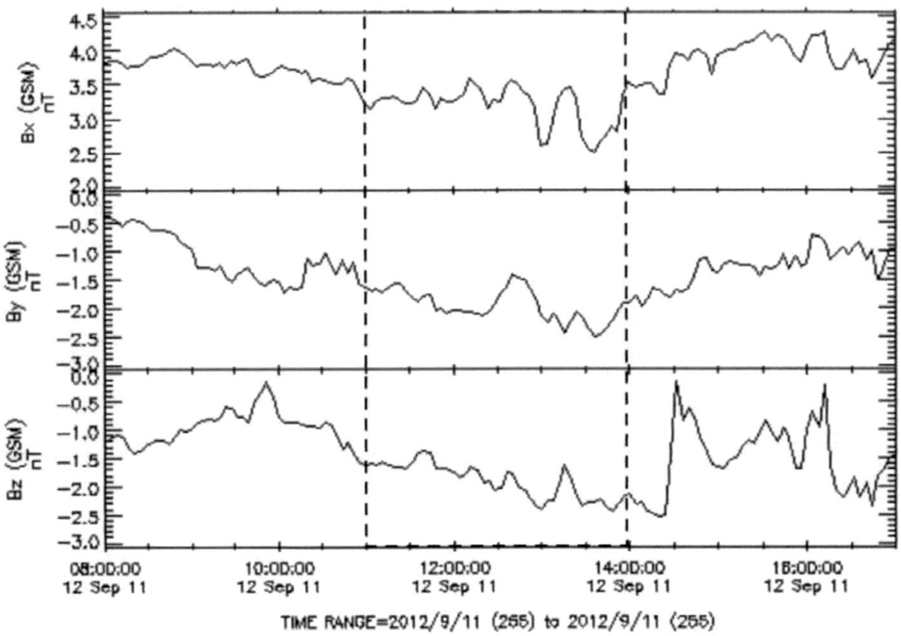

Figure 2. Interplanetary Magnetic Field (IMF) corresponding to 11:00–14:00 Universal Time (UT) on 11th September 2012 (a part surrounded by a broken line), clearly showing a mainly southward magnetic field in the solar wind.

and the z-component of IMF, $B_Z = -1.6$ nT. Subsequently, we used time series of solar wind parameters measured from the ACE satellite to simulate the ionospheric electric potential in the interval of 11:00–14:00 11th September 2012.

Figure 3 shows the simulation results for the interval of 11:00–14:00 11th September 2012. From left to right, the panels show contour maps in a time sequence of UT = 12:19, 12:49 and 13:22 UT. The top row shows the northern hemisphere and the bottom row shows the southern hemisphere. The color contours are representing the electric potential. Thick (thin) contours show positive (negative) electric potential. Each panel is seen from above the North Pole. Noon is to the top and dusk to the left.

In the case of intense ionospheric convection driving, it is well known from previous work that the IMF B_Y introduces dawn-dusk asymmetries in the ionospheric convection pattern ([8], and references therein). In the Northern Hemisphere ionosphere, it was found that a round cell is formed at dawn and a crescent cell at dusk under a negative IMF B_Y condition [9]. The results shown in Figure 3 coincides with many of the above earlier studies.

The global network of Super Dual Auroral Radar Network (SuperDARN) radars measures the plasma drift in the ionosphere in the auroral and polar regions [10]. We compared the ionospheric plasma drift estimated from the global MHD simulation and that observed from SuperDARN radars.

In this paper, we present procedures for deriving the plasma drift from the global MHD simulation. We can get the calculated ionospheric potential from the left panel of Figure 4 and the location where we can get SuperDARN radar data from the right panel of Figure 4. Then we can get the plasma drift from the center panel of Figure 5 and the plasma motion in the line-of-sight direction (LOSV) from the right panel of Figure 5.

The left panel of Figure 4 shows calculated ionospheric potential with color contours. The

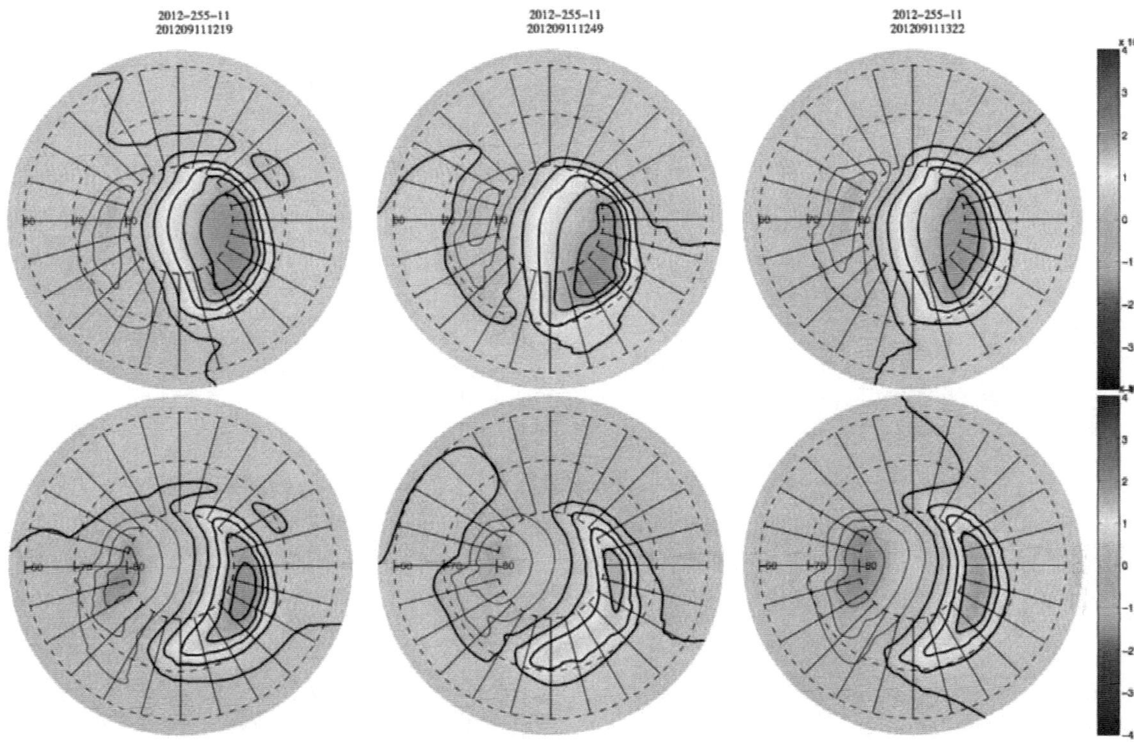

Figure 3. Overview of simulation results for the interval of 11:00–14:00 11th September 2012. The top row shows the northern hemisphere and the bottom row shows the southern hemisphere. The color contours are representing the electric potential. Thick (thin) contours show positive (negative) electric potential. From left to right panels, the panels show snapshots of contour maps at UT = 12:19, 12:49 and 13:22 UT. Each panel is seen from above the North Pole. Noon is to the top and dusk to the left.

right panel of Figure 4 contains the plasma motion in the line-of-sight direction (LOSV) of the radar beams. From the calculated ionospheric potential and the magnetic field model, we calculated $\mathbf{E} \times \mathbf{B}$ plasma drifts and regarded them as plasma convections in the ionosphere. Arrows in the left panel of Figure 4 means the reproduced ionospheric convection.

Figure 5 shows the results at 11:12 UT 11th September 2012 in the northern hemisphere. The panels are in the same format as Figure 4. The simulated ionospheric convection velocities are projected to the center panel as the plasma motion at the point where we can get SuperDARN radar data (shown as arrows). Then we can calculate LOSVs of the radar beams from the simulated ionospheric convection. Arrows in the right panel mean the calculated LOSVs of the radar beams (shown as arrows).

Herewith we can compare the observed and calculated LOSVs at the point where we can get radar data. Figure 6 shows the time variation of the calculated and observed LOSVs at the King Salmon radar site. The horizontal axis is time and the vertical axis is the distance from the radar site. The observed LOSVs are shown in the left panels, and the calculated LOSVs are shown in the right panels. From top to bottom rows show the LOSVs from beam 5, 6, 7, and 8, respectively.

From the results of the observed LOSVs, we found that fast (> 700 m/s) westward flows appeared near the radar site at 11:30 to 12:30 UT. The regions where the fast flows appeared

IOP Conf. Series: Journal of Physics: Conf. Series **1035** (2018) 012007 doi:10.1088/1742-6596/1035/1/012007

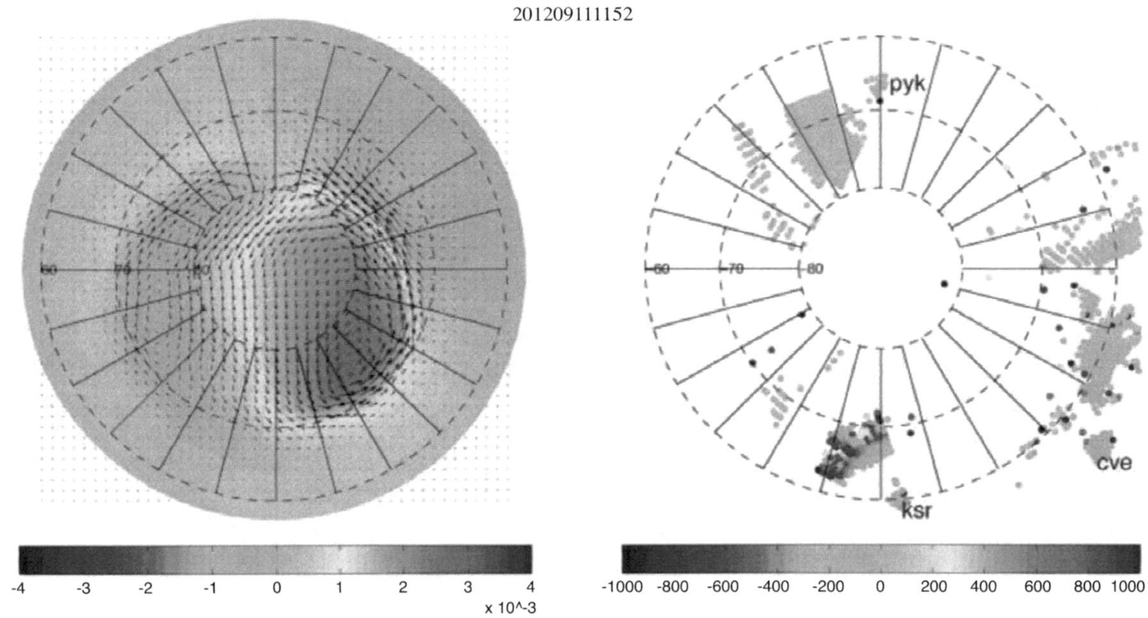

Figure 4. The left panel shows the simulated ionospheric potential (shown as color contours) and the ionospheric convection (shown as arrows). The right panel shows the plasma motion in the line-of-sight direction (LOSV) of the radar beams derived from the Ionospheric scatter echoes observed by SuperDARN radars. This figure shows the results at 11:52 UT 11th September 2012 in the north hemisphere. Strings consists three characters mean radar name (**cve**: Christmas Valley East, **pyk**: Pykkvibaer, and **ksr**: King Salmon).

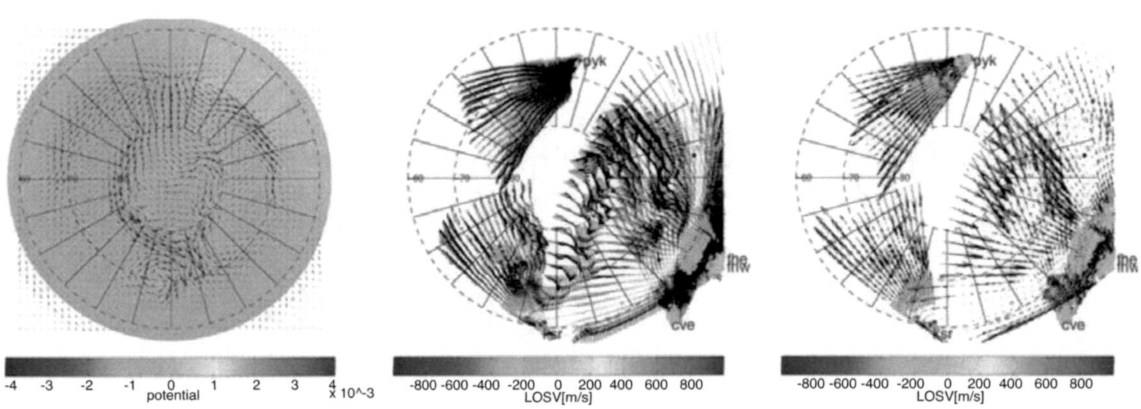

Figure 5. This figure shows the results at 11:12 UT 11th September 2012 in the north hemisphere. These panels are in the same format as Figure 4. The center panel contains the plasma motion (shown as arrows) at the point that we can get SuperDARN radar data. The right panel contains the plasma motion in the line-of-sight direction (shown as arrows) of the radar beams . Strings consists three characters mean radar name (**cve**: Christmas Valley East, **fhe**: Fort Hays East, and **fhw**: Fort Hays West).

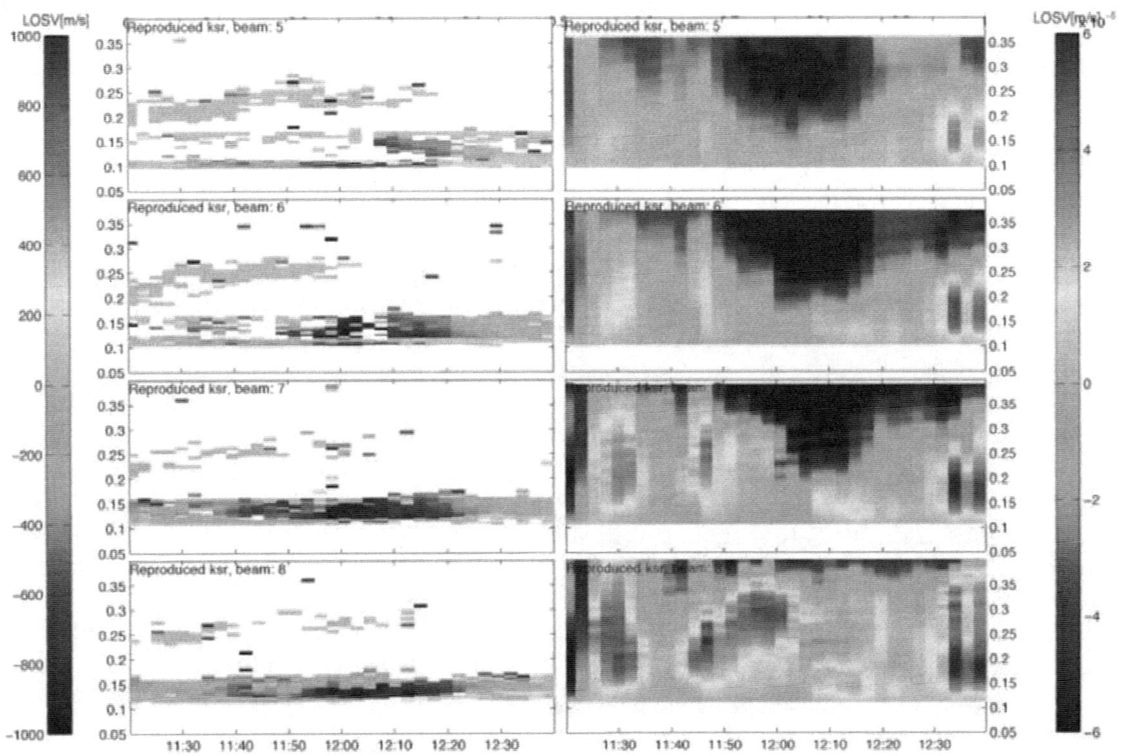

Figure 6. This figure shows the time variation of the calculated and observed LOSVs at the King Salmon radar site. The horizontal axis is time and the vertical axis is the distance from the radar site. The observed LOSVs are shown in the left panels, and the calculated LOSVs are shown in the right panels. From top to bottom rows show the LOSVs from beam 5, 6, 7, and 8, respectively.

correspond to the blue regions in the right panel of Figure 4. From the data in Figure 4, is can be seen that the fast flows expand to lower latitudes ($60° \sim 70°$). Satellites often show significant westward convections in the dusk subauroral region. The rapid westward stream is distinguished with the two-cell convection pattern as shown in Figure 3. This kind of events is called Sub Aurora Polarization Stream (SAPS) [11]. We think that this significant fluctuation was caused by SAPS. On the other hand from the calculated LOSVs, we could not found such fluctuations in the same region and in the same time interval.

5. Summary

For global MHD simulations, we chose the data assimilation target to be the plasma convection velocity in the ionosphere. In our simulation model, we use solar wind parameters from the ACE satellite as outer boundary conditions. We reproduced the ionospheric electrical potential and calculated the ionospheric $\mathbf{E} \times \mathbf{B}$ plasma drift to compare the ionosphere plasma convection velocity obtained from simulation data with actual observed values. The observed values were the observed Doppler shift of ionospheric scatter echoes from the SuperDARN radar network. We reproduced the plasma motion in the line-of-sight direction (LOSV) of the radar beams at the point where SuperDARN radar data were obtained.

The results of our analysis revealed that the reproduced dawn-dusk asymmetries in the ionospheric convection pattern coincide with many of the above earlier studies. However, significant fluctuations due to SAPS were not seen in the reproduced LOSVs while measured data values showed LOSVs. It seems possible that the M-I coupling model currently used in global MHD simulations may not be capturing the effects of polarization fields on ionospheric convections or magnetospheric convections. We confirmed the need to design a M-I coupling model that incorporates more realistic processes of interaction between the magnetosphere and ionosphere. One accomplishment of this study is the establishment of an environment for reproducing observed values from global MHD simulation data and comparing observed and calculated values. In future work, we will attempt to establish data assimilation methods that make use of more massive volumes of data, including not only the SuperDARN radar data but also data on field-aligned currents (FACs) measured by the AMPERE satellite.

Acknowledgments

This study was supported by National Institute of Polar Research (NIPR) through General Collaboration Project no. 27–11 The observed doppler shift of ionospheric scatter echoes are provided by IUGONET (Inter-university Upper Atmosphere Global Observation NETwork) project [12–15]. Data files are distributed in the CDF format through the ERG-SC repository `https://ergsc.isee.nagoya-u.ac.jp/index.shtml.en`. We thank to the National Institute of Information and Communications Technology (NICT), Japan, Dartmouth College, USA, and University of Leicester, UK for providing SuperDARN radar data from King Salmon, Christmas Valley East, and Pykkvibaer. We also thank to WDC-C2 Kyoto AE index service, AE stations and the persons who derive the index. The authors would like to thank National Space Science Data Center for providing the ACE data used in this study.

References

[1] Lyon J, Brecht S H, Fedder J A and Palmadesso P 1980 *Geophys. Res. Lett.* **7** 721
[2] Ogino T 1986 *J. Geophys. Res.* **91** 6791
[3] Fedder J A and Lyon J G 1995 *J. Geophys. Res.* **100** 3623
[4] Tanaka T 1995 *J. Geophys. Res.* **100** 12057
[5] Tanaka T 2003 *Space Plasma Simulation Lecture Notes in Physics* **615** 275
[6] Tanaka T, Nakamizo A, Yoshikawa A, Fujita S, Shinagawa H, Shimazu H, Kikuchi T and Hashimoto K K 2010 *J. of Geophys. Res.* **115** A05220
[7] Rostoker G 1972 *Rev. Geophys.* **10** 935
[8] Haaland S E, Runov A and Forsyth C 2017 *Dawn-dusk asymmetries in planetary plasma environments* (Hoboken, NJ: John Wiley)
[9] Lockwood M 1991 Modelling the high-latitude ionosphere *Proceedings IEE Colloqium on 'National Radio Propagation Programme'* **10** (London, UK: IEE)
[10] Greenwald R A, Baker K B, Dudeney J R, Pinnock M, Jones T B, Thomas E C, Villain J P, Cerisier J C, Senior C, Hanuise C, Hunsucker R D, Sofko G, Koehler J, Nielsen E, Pellinen R, Walker A D M, Sato N and Yamagishi H, 1995 *Space Sci. Rev.* **71** 761
[11] Erickson P J, Matsui H, Foster J C, Torbert R B, Ergun R E, Ergun, Khotyaintsev Yu V, Lindqvist PA, Argall M R, Farrugia C J, Paulson K W, Strangeway R J and Magnes W 2016 *Geophys. Res. Lett.* **43** 7294
[12] Yatagai A, Tanaka , Abe S, Shinbori A, Yagi M, UeNo S, Koyama Y, Umemura N, Nosé M, Hori T, Sato Y, Hashiguchi N O, Kaneda N and IUGONET project team 2014 *Data Sci. J.* **13** PDA37
[13] Hayashi H, Koyama Y, Hori T, Tanaka Y, Abe S, Shinbori A, Kagitani M, Kouno T, Yoshida D, UeNo S, Kaneda N, Yoneda M, Umemura N, Tadokoro H, Motoba T and IUGONET project team 2013 *Data Sci. J.* **12** WDS179
[14] Abe S, Umemura N, Koyama Y, Tanaka Y, Yagi M, Yatagai A, Shinbori A, UeNo S, Sato Y and Kaneda N 2014 *Earth, Planets and Space* **66** 133
[15] Tanaka Y, Shinbori A, Hori T, Koyama Y, Abe S, Umemura N, Sato Y, Yagi M, Ueno S, Yatagai A, Ogawa Y and Miyoshi Y 2013 *Adv. Polar Sci.* **24** 231

IOP Conf. Series: Journal of Physics: Conf. Series **1035** (2018) 012008 doi:10.1088/1742-6596/1035/1/012008

A Study of Steady Flow of a Polyatomic Gas by Extended Thermodynamics with Six Fields

Masaru Sugiyama

Nagoya Institute of Technology, Nagoya 466-8555, Japan

E-mail: sugiyama@nitech.ac.jp

Abstract. Steady flow of a polyatomic gas is studied by the extended thermodynamic theory with six independent fields (ET_6). It is shown that the ET_6 theory can describe the effects on a flow due to the relaxation process of the internal motion such as molecular rotation and vibration, which can not be predicted by the Euler perfect fluid theory. And thereby characteristic features of the ET_6 theory itself can be elucidated.

1. Introduction

Extended thermodynamics (ET) [1, 2, 3] is a theory for studying highly nonequilibrium phenomena, which are out of applicability range of thermodynamics of irreversible processes (TIP) with the local equilibrium assumption [4]. The ET theory of *rarefied monatomic gases* [1, 2, 5] adopts the same hierarchy structure of the balance equations as the moment theory based on the Boltzmann equation [6] such as Grad's moment theory [7]. While the ET theory of *rarefied polyatomic gases* has been developed [3, 8, 9, 10, 11, 12, 13], where the binary hierarchy of the balance equations including the dynamic pressure (nonequilibrium pressure) as an independent field is adopted. The theory, in particular the theory with 14 independent fields, has been successfully confirmed by comparing the theoretical predictions to the experimental data of linear waves [14], shock waves [15], and light scattering [16] in the range where the TIP theory (in this case, the Navier-Stokes and Fourier theory) is invalid.

The role of the dynamic pressure can be highlighted in the ET theory with 6 fields (ET_6): mass density, velocity, specific internal energy, and dynamic pressure [17, 18, 19]. The correspondence relation between ET_6 and the Meixner theory of relaxation processes [20, 21] was established in [17, 22]. The theory with a nonlinear constitutive equation was also proposed and developed [22, 23, 24, 25]. Furthermore the ET theory with two relaxation processes was studied by introducing one more independent field [26].

Recently an ET theory of *dense polyatomic gases* with 6 independent fields was proposed by postulating a principle of duality between rarefied and dense gases in the paper [27] where two kinds of nonequilibrium temperature are introduced: One is related to the subsystem with the kinetic and potential energies and the other is related to the subsystem of the internal modes such as molecular rotation and vibration. The importance of the internal relaxation process due to the molecular energy exchange between two subsystems for understanding the nature of the dynamic pressure was also emphasized [27].

The purpose of the present paper is to study a steady flow of a polyatomic gas by the ET_6 theory. It is shown that, in contrast to the Euler perfect fluid theory, the ET_6 theory predicts

IOP Conf. Series: Journal of Physics: Conf. Series **1035** (2018) 012008 doi:10.1088/1742-6596/1035/1/012008

some remarkable effects on a flow due to the relaxation process of the internal molecular modes. And thereby characteristic features of the ET_6 theory itself can be elucidated.

2. Brief summary of the ET_6 theory of dense polyatomic gases

In this section, for the sake of completeness, we summarize briefly the ET_6 theory [27]. We study a gas with the Hamiltonian in the form:

$$H = H^{K+U} + H^I,$$

where H^{K+U} $(= K + U)$ is the Hamiltonian of the subsystem having the kinetic energy K of molecular translational motion and the potential energy U among molecules, and H^I is the Hamiltonian of the subsystem of the internal motion of constituent molecules such as molecular rotation and vibration.

In an equilibrium state with the mass density ρ and the temperature T, the pressure p can be divided into two parts: the pressure p^{K+U} due to the $(K + U)$-subsystem and the pressure p^I due to the I-subsystem. However, as $p^I = 0$, we have the thermal equation of state:

$$p = p(\rho, T) \equiv p^{K+U}(\rho, T).$$

Similarly we have the caloric equation of state for the specific internal energy ε:

$$\varepsilon = \varepsilon(\rho, T) \equiv \varepsilon^{K+U}(\rho, T) + \varepsilon^I(T),$$

where ε^{K+U} and ε^I are, respectively, the specific internal energies of the $(K+U)$-subsystem and of the I-subsystem. Note that ε^I depends only on T. The specific entropy density s is obtained in the following form [28]:

$$s = s^{K+U}(\rho, T) + s^I(T), \tag{1}$$

where s^{K+U} and s^I are the specific entropy densities of the two subsystems. The Gibbs relations of the subsystems are expressed as

$$\begin{aligned} T\mathrm{d}s^{K+U} &= \mathrm{d}\varepsilon^{K+U} - \frac{p}{\rho^2}\mathrm{d}\rho, \\ T\mathrm{d}s^I &= \mathrm{d}\varepsilon^I. \end{aligned} \tag{2}$$

In the ET_6 theory, a nonequilibrium state is specified by the quantities $\{\rho, v_i, \vartheta, \Theta\}$, where v_i is the velocity, θ and Θ are the nonequilibrium temperatures of the $(K + U)$-subsystem and of the I-subsystem, respectively [27]. Then the time-evolution of a gas is governed by the following system of balance equations:

$$\begin{aligned} &\frac{\partial \rho}{\partial t} + \frac{\partial}{\partial x_i}(\rho v_i) = 0, \\ &\frac{\partial \rho v_j}{\partial t} + \frac{\partial}{\partial x_i}\{p(\rho, \vartheta)\delta_{ij} + \rho v_i v_j\} = 0, \\ &\frac{\partial}{\partial t}\{2\rho\varepsilon^{K+U}(\rho, \vartheta) + 2\rho\varepsilon^I(\Theta) + \rho v^2\} + \\ &\quad + \frac{\partial}{\partial x_i}\{[2p(\rho, \vartheta) + 2\rho\varepsilon^{K+U}(\rho, \vartheta) + 2\rho\varepsilon^I(\Theta) + \rho v^2]\, v_i\} = 0, \\ &\frac{\partial}{\partial t}\{2\rho\varepsilon^I(\Theta)\} + \frac{\partial}{\partial x_i}\{2\rho\varepsilon^I(\Theta)v_i\} = -\hat{P}_{ll}, \end{aligned} \tag{3}$$

where $-\hat{P}_{ll}$ is the production term. The first three equations are conservation laws of mass, momentum, and energy. The last equation expresses the irreversible energy exchange between the two subsystems.

The entropy balance equation is given by

$$\frac{\partial h}{\partial t} + \frac{\partial h_i}{\partial x_i} = \Sigma \geqslant 0, \tag{4}$$

where h $(= \rho\eta)$ is the entropy density, h_i is the entropy flux, and Σ is the entropy production. In the ET$_6$ theory, we can prove $h_i = hv_i$. The nonequilibrium specific entropy density η is given as follows:

$$\eta = \eta(\rho, \vartheta, \Theta) = s^{K+U}(\rho, \vartheta) + s^I(\Theta). \tag{5}$$

If $\vartheta = \Theta = T$ with T being the (local) equilibrium temperature, equation (5) coincides with (1).

3. Basic equations of a steady flow

From (3), a steady flow is governed by the following system of equations:

$$\begin{aligned}
&\frac{\partial}{\partial x_i}(\rho v_i) = 0, \\
&\frac{\partial p(\rho,\vartheta)}{\partial x_j} + \rho v_i \frac{\partial v_j}{\partial x_i} = 0, \\
&v_i \frac{\partial}{\partial x_i}\left(2\frac{p(\rho,\vartheta)}{\rho} + 2\varepsilon^{K+U}(\rho,\vartheta) + 2\varepsilon^I(\Theta) + v^2\right) = 0, \\
&2\rho v_i \frac{\partial \varepsilon^I(\Theta)}{\partial x_i} = -\hat{P}_{ll}.
\end{aligned} \tag{6}$$

And the entropy balance equation (4) now reduces to

$$\rho v_i \frac{\partial \eta}{\partial x_i} = \Sigma \geqslant 0. \tag{7}$$

Let us pay attention to the variation of quantities along a streamline. From $(6)_{2,3,4}$, we have

$$\begin{aligned}
&\delta p(\rho,\vartheta) + \rho v \delta v = 0, \\
&v\delta\left(2\frac{p(\rho,\vartheta)}{\rho} + 2\varepsilon^{K+U}(\rho,\vartheta) + 2\varepsilon^I(\Theta) + v^2\right) = 0, \\
&2\rho v \delta\varepsilon^I(\Theta) = -\hat{P}_{ll},
\end{aligned} \tag{8}$$

where

$$\delta \equiv \frac{v_i}{v}\frac{\partial}{\partial x_i} \qquad (v \equiv \sqrt{v_i v_i})$$

is the directional derivative along a streamline. And, from (7), we have

$$\rho v \delta\eta = \Sigma. \tag{9}$$

Equations (8) are rewritten in terms of $\{\delta v, \delta\rho, \delta\vartheta, \delta\Theta\}$ as follows:

$$\begin{aligned}
&\rho v\delta v + p_\rho(\rho,\vartheta)\delta\rho + p_\vartheta(\rho,\vartheta)\delta\vartheta = 0, \\
&v\delta v + \left(-\frac{p(\rho,\vartheta)}{\rho^2} + \frac{p_\rho(\rho,\vartheta)}{\rho} + \varepsilon_\rho^{K+U}(\rho,\vartheta)\right)\delta\rho + \left(\frac{p_\vartheta(\rho,\vartheta)}{\rho} + c_v^{K+U}(\rho,\vartheta)\right)\delta\vartheta + c_v^I(\Theta)\delta\Theta = 0, \\
&2\rho v c_v^I(\Theta)\delta\Theta = -\hat{P}_{ll},
\end{aligned} \tag{10}$$

where the suffix indicates the partial derivative, for example, $p_\rho(\rho,\vartheta) \equiv \partial p(\rho,\vartheta)/\partial\rho$, and c_v^{K+U} and c_v^I are the specific heats of the subsystems defined by

$$c_v^{K+U}(\rho,\vartheta) = \frac{\partial\varepsilon^{K+U}(\rho,\vartheta)}{\partial\vartheta}, \qquad c_v^I(\Theta) = \frac{d\varepsilon^I(\Theta)}{d\Theta}.$$

These specific heats are positive, and the total specific heat is given by $c_v = c_v^{K+U} + c_v^I$.

4. Discussions

From $(10)_{1,2}$, the variation of the mass flux $\delta(\rho v)$ is expressed in terms of δv and $\delta\Theta$ as follows:

$$\delta(\rho v) = \rho\left(1 - \frac{v^2}{u^2}\right)\delta v + \frac{v}{u^2}\frac{p_\vartheta(\rho,\vartheta)c_v^I(\Theta)}{c_v^{K+U}(\rho,\vartheta)}\delta\Theta, \tag{11}$$

where u is the characteristic velocity [27] given by

$$\begin{aligned} u^2 &= p_\rho(\rho,\vartheta) + \frac{p_\vartheta(\rho,\vartheta)}{c_v^{K+U}(\rho,\vartheta)}\left\{\frac{p^{K+U}(\rho,\vartheta)}{\rho^2} - \varepsilon_\rho^{K+U}(\rho,\vartheta)\right\} \\ &= p_\rho(\rho,\vartheta) + \frac{\vartheta p_\vartheta^2(\rho,\vartheta)}{\rho^2 c_v^{K+U}(\rho,\vartheta)}. \end{aligned}$$

In particular, in an equilibrium case with the condition $\vartheta = \Theta = T$, the velocity u is given by

$$u_E^2 = p_\rho(\rho,T) + \frac{Tp_T^2(\rho,T)}{\rho^2 c_v^{K+U}(\rho,T)}.$$

It is interesting to remind that the counterpart of equation (11) in the Euler theory is given by

$$\delta(\rho v) = \rho\left(1 - \frac{v^2}{u_{Euler}^2}\right)\delta v, \tag{12}$$

where u_{Euler} is the sound velocity given by

$$u_{Euler}^2 = p_\rho(\rho,T) + \frac{Tp_T^2(\rho,T)}{\rho^2 c_v(\rho,T)}.$$

Because the system of ET_6 includes the Euler system as the subsystem [29], we have the following *subcharacteristic condition*:

$$u_E{}^2 - u_{Euler}^2 = \frac{Tp_T^2(\rho,T)}{\rho^2}\frac{c_v^I(T)}{c_v(\rho,T)c_v^{K+U}(\rho,T)} > 0.$$

Other interesting relations along a streamline are as follows: From $(10)_{1,2}$, we have

$$\delta\varepsilon = \frac{p(\rho,\vartheta)}{\rho^2}\delta\rho \tag{13}$$

and, from (5), (2), (13) and (9), we have

$$\delta\eta = c_v^I(\Theta)\left(\frac{1}{\Theta} - \frac{1}{\vartheta}\right)\delta\Theta = \frac{\Sigma}{\rho v}. \tag{14}$$

Some noticeable points are summarized as follows:

(i) In the Euler theory, as seen from (12), we notice the well-known result that the mass flux ρv reaches the maximum value at the place where the velocity v is equal to the local velocity of sound u_{Euler} [30, 31].

While, in the ET$_6$ theory, from (11), mass flux attains the maximum value at the place where the following condition is satisfied:

$$\rho\left(1-\frac{v^2}{u^2}\right)\delta v = -\frac{v}{u^2}\frac{p_\vartheta(\rho,\vartheta)c_v^I(\Theta)}{c_v^{K+U}(\rho,\vartheta)}\delta\Theta.$$

It is interesting to note that the sign of $\delta v/\delta\Theta$ depends on the condition whether v is greater than the local characteristic speed u or not.

(ii) At the place where the condition $v = u$ is fulfilled, we have the relation:

$$\delta(\rho v) = \frac{1}{u}\frac{p_\vartheta(\rho,\vartheta)c_v^I(\Theta)}{c_v^{K+U}(\rho,\vartheta)}\delta\Theta = \frac{1}{u^2}\frac{p_\vartheta(\rho,\vartheta)}{\rho c_v^{K+U}(\rho,\vartheta)}\frac{\Sigma}{\left(\frac{1}{\Theta}-\frac{1}{\vartheta}\right)}.$$

Therefore, if $p_\vartheta > 0$, the quantities $\delta(\rho v)$, $\delta\Theta$, and $\vartheta - \Theta$ have the same sign.

(iii) When we take the monatomic gas limit by letting ε^I tend to zero, it is easy to see that the relation (11) reduces to (12).

(iv) Let us study a steady flow in a nozzle parallel to the x-direction. We adopt the approximation of a quasi-one-dimensional flow. Then, in addition to the system of equations (8), we have an equation that comes from the conservation law of mass:

$$\delta(\rho v A) = 0, \tag{15}$$

where A is the cross section of a nozzle, and δ in this case is

$$\delta \equiv \frac{d}{dx}.$$

From (11) and (15), we have the relation:

$$\frac{1}{v}\left(1-\frac{v^2}{u^2}\right)\delta v = -\frac{1}{A}\delta A - \frac{1}{u^2}\frac{p_\vartheta(\rho,\vartheta)c_v^I(\Theta)}{\rho c_v^{K+U}(\rho,\vartheta)}\delta\Theta = -\frac{1}{A}\delta A - \frac{1}{u^2}\frac{p_\vartheta(\rho,\vartheta)}{\rho^2 c_v^{K+U}(\rho,\vartheta)}\frac{\Sigma}{\left(\frac{1}{\Theta}-\frac{1}{\vartheta}\right)}. \tag{16}$$

While, in the Euler theory, we have the following relation from (12) and (15):

$$\frac{1}{v}\left(1-\frac{v^2}{u^2_{Euler}}\right)\delta v = -\frac{1}{A}\delta A.$$

From the relation (16), we understand that the internal molecular relaxation process gives a new effect on a flow in addition to the effect of the variation of the cross section.

(v) From the relation (14), we notice that the quantities $\delta\Theta$ and $\vartheta - \Theta$ have the same sign at all places on a streamline.

The above qualitative analysis is valid for any production term $\hat{P}_{ll}$, provided that it is compatible with the entropy principle ($\Sigma > 0$). While in the case of a quantitative analysis, for example numerical analysis, the choice of an explicit expression for $\hat{P}_{ll}$ is required. Its simplest expression may be given by [27]

$$\hat{P}_{ll} = \frac{2\rho}{\tau_\Delta}\Delta,$$

where $\tau_\Delta(\rho, T)(> 0)$ is the relaxation time for the energy exchange between two subsystems:

$$\Delta = \varepsilon^I(\Theta) - \varepsilon^I(T).$$

The final remarks are as follows: i) In a rarefied-gas limit where the intermolecular potential U can be neglected, we have more explicit expressions of the relations obtained above. ii) It is interesting to study real-gas effects by adopting, for example, van der Waals equations of state. iii) Numerical analysis of typical nozzle flows is necessary to obtain quantitative results. Detailed analysis on these subjects will be reported soon elsewhere.

Acknowledgments

I express sincere thanks to Prof. S. Taniguchi and Prof. T. Arima for organizing "The Meeting on Nonequilibrium Thermodynamics and Statistical Physics: From Rational Modeling to Its Applications" held in Fukuoka 2017.

References

[1] Müller I and Ruggeri T 1993 *Extended Thermodynamics* 1st ed. (New York: Springer)
[2] Müller I and Ruggeri T 1998 *Rational Extended Thermodynamics* 2nd ed. (New York: Springer)
[3] Ruggeri T and Sugiyama M 2015 *Rational Extended Thermodynamics beyond the Monatomic Gas* (Cham, Heidelbergh, New York, Dorderecht, London: Springer)
[4] de Groot S R and Mazur P 1984 *Non-Equilibrium Thermodynamics* (New York: Dover)
[5] Liu I-S and Müller I 1983 *Arch. Ration. Mech. Anal.* **83** 285
[6] Chapman S and Cowling T G 1970 *The Mathematical Theory of Non-Uniform Gases* (Cambridge: Cambridge University Press)
[7] Grad H 1949 *Comm. Pure Appl. Math.* **2** 331
[8] Arima T, Taniguchi S, Ruggeri T and Sugiyama M 2011 *Continuum Mech. Thermodyn.* **24** 271
[9] Arima T, Taniguchi S, Ruggeri T and Sugiyama M 2013 *Phys. Lett. A* **377** 2136
[10] Arima T and Sugiyama M 2013 *Atti Accad. Pelorit. Pericol.* **91 (S1)** A1
[11] Ruggeri T and Sugiyama M 2014 *Acta Appl. Math.* **132** 527
[12] Arima T, Mentrelli A and Ruggeri T 2014 *Ann. Phys.* **345** 111
[13] Arima T, Mentrelli A and Ruggeri T 2014 *Rend. Lincei Mat. Appl.* **25** 275
[14] Arima T, Taniguchi S, Ruggeri T and Sugiyama M 2013 *Continuum Mech. Thermodyn.* **25** 727
[15] Taniguchi S, Arima T, Ruggeri T and Sugiyama M 2014 *Phys. Rev. E* **89** 013025
[16] Arima T, Taniguchi S and Sugiyama M 2013 *Proc. 34th Symp. Ultrasonic Electronics* 15
[17] Arima T, Taniguchi S, Ruggeri T and Sugiyama M 2012 *Phys. Lett. A* **376** 2799
[18] Arima T, Ruggeri T, Sugiyama M and Taniguchi S 2014 *Meccanica* **49** 2181
[19] Taniguchi S, Arima T, Ruggeri T and Sugiyama M 2014 *Phys. Fluids* **26** 016103
[20] Meixner J 1943 *Ann. Physik* **43** 470
[21] Meixner J 1952 *Acoustica* **2** 101
[22] Arima T, Ruggeri T, Sugiyama M and Taniguchi S 2015 *Int. J. Non-Linear Mech.* **72** 6
[23] Taniguchi S, Arima T, Ruggeri T and Sugiyama M 2016 *Int. J. Non-Linear Mech.* **79** 66
[24] Arima T, Ruggeri T, Sugiyama M and Taniguchi S 2016 *Ricerche mat.* **65** 263
[25] Taniguchi S, Arima T, Ruggeri T and Sugiyama M 2016 *Ricerche mat.* **65** 279
[26] Arima T, Ruggeri T and Sugiyama M 2017 *Phys. Rev. E* **96** 042143
[27] Arima T, Ruggeri T and Sugiyama M 2017 *Phys. Rev. Fluids* **2** 013401
[28] Landau L D and Lifshitz E M 1980 *Statistical Physics* (Oxford: Pergamon Press)
[29] Boillat G and Ruggeri T 1997 *Arch. Rational Mech. Anal.* **137** 305
[30] Landau L D and Lifshitz E M 1959 *Fluid Mechanics* (Oxford: Pergamon Press)
[31] Liepmann H W and Roshko A 1960 *Elements of Gas Dynamics* (London: John Wiley and Sons)

IOP Conf. Series: Journal of Physics: Conf. Series **1035** (2018) 012009 doi:10.1088/1742-6596/1035/1/012009

Shock wave structure in rarefied polyatomic gases with large relaxation time for the dynamic pressure

Shigeru Taniguchi[1], Takashi Arima[2], Tommaso Ruggeri[3] and Masaru Sugiyama[4]

[1] Department of Creative Engineering, National Institute of Technology, Kitakyushu College, Kitakyushu, Japan

[2] Department of Mechanical Engineering, Faculty of Engineering, Kanagawa University, Yokohama, Japan

[3] Department of Mathematics and Alma Mater Research Center on Applied Mathematics AM[2], University of Bologna, Bologna, Italy

[4] Graduate School of Engineering, Nagoya Institute of Technology, Nagoya, Japan

E-mail: taniguchi.shigeru@kct.ac.jp, arima@kanagawa-u.ac.jp, tommaso.ruggeri@unibo.it, sugiyama@nitech.ac.jp

Abstract. The shock wave structure in rarefied polyatomic gases is analyzed based on extended thermodynamics (ET). In particular, the case with large relaxation time for the dynamic pressure, which corresponds to large bulk viscosity, is considered by adopting the simplest version of extended thermodynamics with only 6 independent fields (ET_6); the mass density, the velocity, the temperature and the dynamic pressure. Recently, the validity of the theoretical predictions by ET was confirmed by the numerical analysis based on the kinetic theory in [S Kosuge and K Aoki: Phys. Rev. Fluids, Vol. 3, 023401 (2018)]. It was shown that numerical results using the polyatomic version of ellipsoidal statistical model agree with the theoretical predictions by ET for small or moderately large Mach numbers. In the present paper, first, we compare the theoretical predictions by ET_6 with the ones by kinetic theory for large Mach number under the same assumptions, that is, the gas is polytropic and the bulk viscosity is proportional to the temperature. Second, the shock wave structure for large Mach number in a non-polytropic gas is analyzed with the particular interest in the effect of the temperature dependence of specific heat and the bulk viscosity on the shock wave structure. Through the analysis of the case of a rarefied carbon dioxide (CO_2) gas, it is shown that these temperature dependences play important roles in the precise analysis of the structure for strong shock waves.

1. Introduction

In polyatomic gases, in addition to the translational modes, the rotational and vibrational modes (microscopic internal modes) play important roles. In particular, in some temperature range, the internal modes are partially excited and relax much more slowly than the translational modes. Due to the presence of the slowly relaxing modes, the shock wave structure in rarefied polyatomic gas shows different features from the ones in rarefied monatomic gases [1, 2]. The thickness of a shock wave is several orders larger than the mean free path, which is typical length of the thickness in the case of rarefied monatomic gases. Moreover, as the Mach number increases from unity, the profile of the shock wave structure changes from the nearly symmetric profile (Type A) to the asymmetric profile (Type B), and then changes further to the profile composed of

IOP Conf. Series: Journal of Physics: Conf. Series **1035** (2018) 012009 doi:10.1088/1742-6596/1035/1/012009

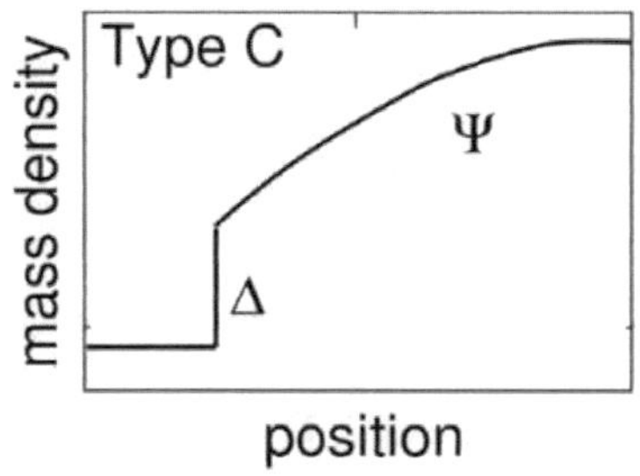

Figure 1. Schematic representation of the mass density profile for the shock wave structure of Type C. The shock wave structure is composed of the thin layer Δ and the thick layer Ψ.

thin and thick layers (Type C) [3–8]. Schematic profile of the mass density of the shock wave structure of Type C is shown in Figure 1. The previous approaches by Gilbarg & Paolucci [9] based on the Navier-Stokes, Fourier equations and by Bethe & Teller [10] based on the Meixner theory [11, 12] are not able to explain these features in a unified way. A unified explanation of these peculiar shock wave structures has remained a long-standing problem. Although the experimental results were obtained in the 1950s, only recently, unified explanation was made.

A promising phenomenological approach is the one based on extended thermodynamics (ET) theory [13–15]. It was shown that the features of the shock wave structure are explained in a unified way by ET with 14 independent variables (ET_{14}); the mass density, the velocity, the temperature, the dynamic pressure (nonequilibrium pressure), the shear stress and the heat flux [16, 17]. The agreement between the theoretical predictions and the experimental data for small or moderately large Mach numbers is quite good [16, 17]. Moreover, it was shown that, when the relaxation time for the dynamic pressure is much larger than other relaxation times, the shock wave structure can be explained satisfactorily [18] based on the simplified ET theory with only 6 independent fields (ET_6) [19, 20]; the mass density ρ, the velocity $\mathbf{v} \equiv (v_i)$, the temperature T and the dynamic pressure Π. The nonlinear version ET_6 theory was also derived only from the general principles without adopting near equilibrium assumption and therefore nonlinear of ET_6 is valid even far from equilibrium [21]. It was shown that the predictions of the shock wave structure based on linear and nonlinear ET_6 theories are substantially the same for small or moderately large Mach numbers and the only small difference appears even for large Mach numbers [22]. Therefore, we expect to analyze safely the structure of a strong shock wave for large Mach numbers by using ET_6.

After the successful explanation about the shock wave structure by ET theories, the confirmation of the results of ET theories was also made by the numerical analysis of kinetic theory. Previously, the approaches based on the Boltzmann equation such as the Direct Simulation Monte Carlo (DSMC) method [23], the Chapman-Enskog method [24] and the moment method [25–28]) has encountered the difficulty in an appropriate modeling of the collision term due to the complexity of the collisional processes. Very recently, it was shown that the polyatomic version of the ellipsoidal statistical (ES) model can explain the features of the shock wave structure [29, 30] by adopting large bulk viscosity, which corresponds to the large relaxation time for the dynamic pressure (See the equation (2) in the next section). The prediction by kinetic theory agrees quite well with the one by ET_{14} for moderately large Mach number; $M_0 = 1.47$, where M_0 represents the unperturbed Mach number. In these papers [29,30], systematic numerical computations on the shock wave structure were carried out even for strong shock waves.

The purpose of the present paper is to make clear the structure of a strong shock wave in rarefied polyatomic gases in which the relaxation time for the dynamic pressure is much larger than other relaxation times. First, we compare the theoretical predictions by ET_6 with the ones obtained by the numerical calculations based on the kinetic theory for large Mach number under the same assumptions, that is, the gas is polytropic and the bulk viscosity is proportional to the temperature. Second, we clarify the effect of the temperature dependence of the specific heat and bulk viscosity on the shock wave structure by considering non-polytropic gases with the general temperature dependence of the bulk viscosity. Through the analysis of the case of a rarefied CO_2 gas, we show the important role of these temperature dependences in the analysis of the structure of strong shock waves.

2. Field equations of ET_6

Let us consider plane shock waves propagating in x direction such that the velocity is expressed as $\mathbf{v} \equiv (v, 0, 0)$. The field equations of ET_6 are summarized in the following balance form with the independent field variables $\mathbf{u} = (\rho, v, T, \Pi)$ [19, 21, 22, 31] :

$$\begin{aligned}
&\frac{\partial \rho}{\partial t} + \frac{\partial}{\partial x}(\rho v) = 0, \\
&\frac{\partial \rho v}{\partial t} + \frac{\partial}{\partial x}\left[(p+\Pi) + \rho v^2\right] = 0, \\
&\frac{\partial}{\partial t}(2\rho\varepsilon + \rho v^2) + \frac{\partial}{\partial x}\left\{\left[2(p+\Pi) + 2\rho\varepsilon + \rho v^2\right] v\right\} = 0, \\
&\frac{\partial}{\partial t}\left[3(p+\Pi) - 2\rho\varepsilon\right] + \frac{\partial}{\partial x}\left\{\left[3(p+\Pi) - 2\rho\varepsilon\right] v\right\} = -\frac{3\Pi}{\tau_\Pi},
\end{aligned} \tag{1}$$

where ε, p and τ_Π are, respectively, the specific internal energy, the (equilibrium) pressure and the relaxation time for the dynamic pressure. In general, the relaxation time for the dynamic pressure τ_Π is a function of the mass density ρ and the temperature T. As it was discussed in the *Introduction*, it was shown that the difference between the theoretical predictions by linear and nonlinear ET_6 theories is not so drastic [22]. For simplicity, we adopt linear ET_6 in the present study. The following caloric and thermal equations of state for rarefied polyatomic gases are adopted:

$$\varepsilon \equiv \varepsilon(T), \qquad p = \frac{k_B}{m}\rho T,$$

where k_B and m are, respectively, the Boltzmann constant and the mass of a molecule. The temperature dependence of ε may be determined by the dependence of the specific heat c_v on the temperature through the following relation:

$$\varepsilon(T) = \frac{k_B}{m}\int_{T_R}^{T} \hat{c}_v(\theta)\,\mathrm{d}\theta$$

with $\hat{c}_v = (m/k_B)c_v$ being the dimensionless specific heat and T_R represents an inessential reference temperature. By using the Maxwellian iteration [13], we obtain the following relation between the relaxation time for the dynamic pressure τ_Π and the bulk viscosity ν:

$$\nu = \left(\frac{2}{3} - \frac{1}{\hat{c}_v}\right) p\tau_\Pi. \tag{2}$$

We can determine the functional forms of $\varepsilon(T)$ and $\tau_\Pi(\rho, T)$ once we have the temperature dependence of $\hat{c}_v$ and the dependence of ν on the mass density and the temperature from the kinetic theoretical consideration and/or experimental data.

It should be noted that the systems of field equations of the ET theory of rarefied polyatomic gases are consistent with kinetic theoretical consideration. The kinetic theory gives the same systems of field equations as the moment equations via Maximum entropy principle for the case of polytropic gases [19, 32–34] and also for non-polytropic gases [35].

3. Shock wave structure

3.1. Settings and numerical method

Due to the Galilean invariance, without loss of generality, we adopt the frame moving with the shock wave such that the shock wave is considered as stationary. It is assumed that both the unperturbed state (the state at $x = -\infty$ before and far from a shock wave) and the perturbed state (the state at $x = \infty$ after and far from a shock wave) are in thermal equilibrium. Hereafter, the quantities with subscript 0 represent the quantities evaluated in the unperturbed state $\mathbf{u}_0 = (\rho_0, v_0, T_0, 0)$ and the quantities with subscript 1 represent the quantities evaluated in the perturbed state $\mathbf{u}_1 = (\rho_1, v_1, T_1, 0)$. The Mach number in the unperturbed state is defined by

$$M_0 \equiv \frac{v_0}{c_0}$$

with c_0 being the sound velocity in the unperturbed state. The relationship between $\mathbf{u}_1$, $\mathbf{u}_0$ and M_0 is given by the Rankine-Hugoniot (RH) conditions for the system of the Euler equations. When the shock velocity is larger than the maximum characteristic velocity, the sub-shock appears [36]. In this case, we incorporate the analytical result on the strength of the sub-shock predicted by the RH condition for the system of ET$_6$ [18]. The algebraic equations obtained by replacing the differentiation in space by the central difference are solved numerically by Newton's method [16, 18, 37].

3.2. Case of a polytropic gas with $\nu \propto T$

In [30], it was shown that theoretical prediction by kinetic theory agrees with the one of ET$_{14}$ for moderately large Mach number $M_0 = 1.47$. In this subsection, we compare the theoretical predictions by ET$_6$ with the ones by kinetic theory shown in [30] for large Mach number under the same assumptions. We assume that $\hat{c}_v$ is independent of the temperature [30] and we introduce the following notation:

$$\hat{c}_v = \frac{D}{2},$$

where $D(> 3)$ is a constant, which represents the degrees of freedom of a molecule. In other words, we consider a polytropic gas in this subsection. In the case of a rarefied monatomic gas, $D = 3$. We also assume that [30]

$$\nu \propto T$$

and therefore the relaxation time τ_Π is independent of temperature and inversely proportional to the mass density.

The following dimensionless quantities are introduced for the analysis based on ET$_6$ [16, 18, 22]:

$$\hat{\rho} \equiv \frac{\rho}{\rho_0}, \quad \hat{v} \equiv \frac{v}{c_0}, \quad \hat{T} \equiv \frac{T}{T_0}, \quad \hat{\Pi} \equiv \frac{\Pi}{\rho_0 \frac{k_B}{m} T_0}, \quad \hat{x} \equiv \frac{x}{\tau_\Pi(\rho_0, T_0) c_0}$$

and for comparison, we also introduce the following dimensionless position x_1 adopted in the numerical analysis of kinetic theory [30]:

$$x_1 \equiv \hat{x} \frac{3(3+\delta)}{2\delta} \sqrt{\frac{\pi}{8} \frac{\delta+5}{\delta+3}} \mathrm{Pr} \frac{\nu}{\mu}$$

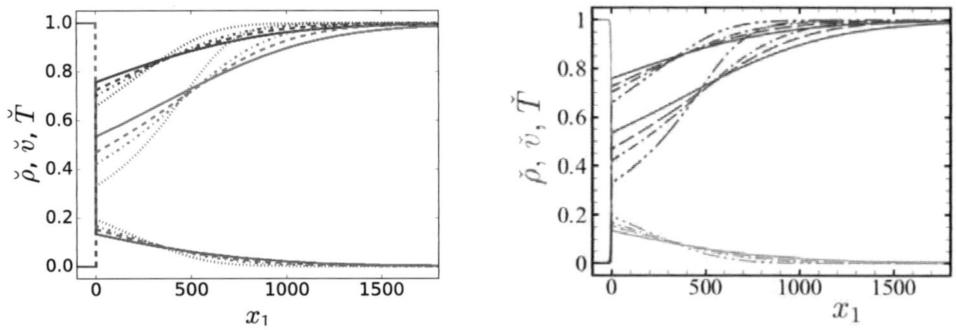

Figure 2. Profiles of the normalized mass density (red curves), normalized velocity (green curves) and normalized temperature (blue curves) predicted by ET_6 (left) for $M_0 = 5$. Four cases with $D = 6$ (solid curves), $D = 7$ (dashed curves), $D = 8$ (dot-dashed curves) and $D = 11$ (dotted curves) are shown. The vertical dashed lines at $x_1 = 0$ represent the jumps of physical quantities across the sub-shock. For comparison, the corresponding theoretical predictions by kinetic theory using ES model (right) are also shown. The right figure is a reproduction of Fig. 6 (a) in [30] (courtesy of K. Aoki).

with Pr and μ being the Prandtl number and the the shear viscosity, respectively and $\delta \equiv D-3$. We adopt the same value as the one used in Fig. 6 of [30]; Pr= 0.761 and $\nu/\mu = 1000$. As the shear stress and the heat flux are not present in ET_6, the values of Pr and μ are used only for converting $\hat{x}$ to x_1. The following normalized quantities are also introduced as it was done in [30]:

$$\breve{\rho} \equiv \frac{\rho - \rho_0}{\rho_1 - \rho_0}, \quad \breve{v} \equiv \frac{v - v_1}{v_0 - v_1}, \quad \breve{T} \equiv \frac{T - T_0}{T_1 - T_0}.$$

Figure 2 shows the profiles of the normalized mass density, normalized velocity and normalized temperature predicted by for $M_0 = 5$. In this Figure, four cases with $D = 6$, $D = 7$, $D = 8$ and $D = 11$ are shown. For comparison, corresponding theoretical predictions by kinetic theory [30] are also shown in Figure 2. We see that all predictions agree with each other quantitatively. The present result strongly supports the validity and the usefulness of both the theoretical predictions by ET_6 and the numerical calculations based on kinetic theory using ES model even for strong shock waves.

3.3. Case of a non-polytropic gas with $\nu \propto T^n$

In the previous subsection, in order to compare the predictions by ET_6 with the one by kinetic theory, we adopt the assumptions of a polytropic gas and $\nu \propto T$. However, the experimental data [38, 39] indicate that these assumptions are not justified in several kinds of rarefied polyatomic gases in some temperature range. In fact, the dependence of $\hat{c}_v$ on the temperature T K for a rarefied carbon dioxide (CO_2) gas is expressed as [22]

$$\hat{c}_v = 1.412 + 8.697 \times 10^{-3}\,T - 6.575 \times 10^{-6}\,T^2 + 1.987 \times 10^{-9}\,T^3 \tag{3}$$

within the third order polynomial by fitting the experimental data [38] in the following temperature range; $220\,\text{K} < T < 1100\,\text{K}$. Moreover, in order to consider the more realistic case, we need to assume the following temperature dependence of ν [9]:

$$\nu \propto T^n,$$

IOP Conf. Series: Journal of Physics: Conf. Series **1035** (2018) 012009 doi:10.1088/1742-6596/1035/1/012009

where the exponent n depends on gases and is determined from the fitting with the experimental data. The value of n is expected to be $n = -1.3$ [22], which is opposite dependence to the case of $\nu \propto T$, through the comparison with the experimental data for a rarefied CO_2 gas [39].

The theoretical predictions of the shock wave structure with some values of n has been reported: The shock wave structure in a rarefied CO_2 gas with $n = 0.935$ was analyzed based on ET by taking non-polytropic effect into account [16,18] and similarly the case of a polytropic gas with $n = 0.935$ was analyzed by kinetic theory [30]. In [22], the shock wave structure in a rarefied CO_2 gas with $n = -1.3$ is analyzed up to $M_0 = 5$. However, the experimental determination of the value of the relaxation time for the dynamic pressure (or the bulk viscosity) is extremely difficult and in this reason a reliable value of n has not yet been known. It was also shown that the shock wave structure based on ET_{14} is not sensitive to the change of n for small or moderately large Mach numbers; $1 < M_0 < 1.47$ [17]. In the present subsection, we adopt $\hat{c}_v$ (3) for a rarefied CO_2 gas and analyze the effect of the value of n on the structure of a strong shock wave.

Let us consider the shock wave structure for $T_0 = 295$ K and $M_0 = 5$. By using RH conditions for the system of Euler equations, we have $T_1 = 1098$ K. The dimensionless specific heat changes from $\hat{c}_v(T_0) = 3.46$ to $\hat{c}_v(T_1) = 5.66$ and we conclude that the change of $\hat{c}_v$ across the shock wave is not negligible. The non-polytropic effect should be considered for precise calculations on the shock wave structure for $M_0 = 5$. As a typical example, in Figure 3, we show the profiles of the dimensionless mass density, velocity, temperature and dynamic pressure for $M_0 = 5$ and $n = 1, 0, -1$. The thin layer of the shock wave structure of Type C is expressed by a sub-shock within the limited resolution of ET_6 [18] and the strength of a sub-shock is determined by RH conditions for the system of ET_6, which are derived from the left-hand side of (1). Therefore the strength of the sub-shock is independent of the values of n and the difference between the predictions with the different values of n appears only in the thick-layer region. Even though, we see the large difference between predictions with different n. The profiles in the thick layer strongly depend on the value of n. It is concluded that the temperature dependences of the specific heat and the bulk viscosity become important in the analysis of strong shock waves for large Mach numbers such as $M_0 = 5$.

4. Summary and concluding remarks

In the present paper, we have analyzed the shock wave structure on the basis of extended thermodynamics (ET) of rarefied polyatomic gases in which the relaxation time for the dynamic pressure is much larger than other relaxation times. We have adopted the ET theory with 6 independent fields (ET_6) and have shown the usefulness of ET_6 through the comparison with the numerical results obtained from the kinetic theory under the same assumptions, that is, the gas is polytropic and the bulk viscosity is proportional to the temperature. It has also been shown that the temperature dependences of the specific heat and the bulk viscosity strongly affect the structure of strong shock waves. The phenomenological approach based on ET enables us to incorporate these temperature dependences in a consistent way.

Lastly, we make the following remarks:

(I) In the present analysis, the simplest extended thermodynamics theory with only 6 independent variables is adopted. Although the thin layer of the shock wave structure of Type C is obtained as the sub-shock within the resolution of ET_6, it is shown that ET with 14 independent variables can describe the fine structure in the thin layer at least for moderate Mach numbers [16]. Detailed analysis of the fine structure by using ET with more independent variables are the future subject.

(II) In the case of ET_6, there exists only one characteristic velocity propagating in the positive direction and the characteristic value of the shock velocity for the sub-shock formation is determined analytically. However, ET theories with more independent variables have several

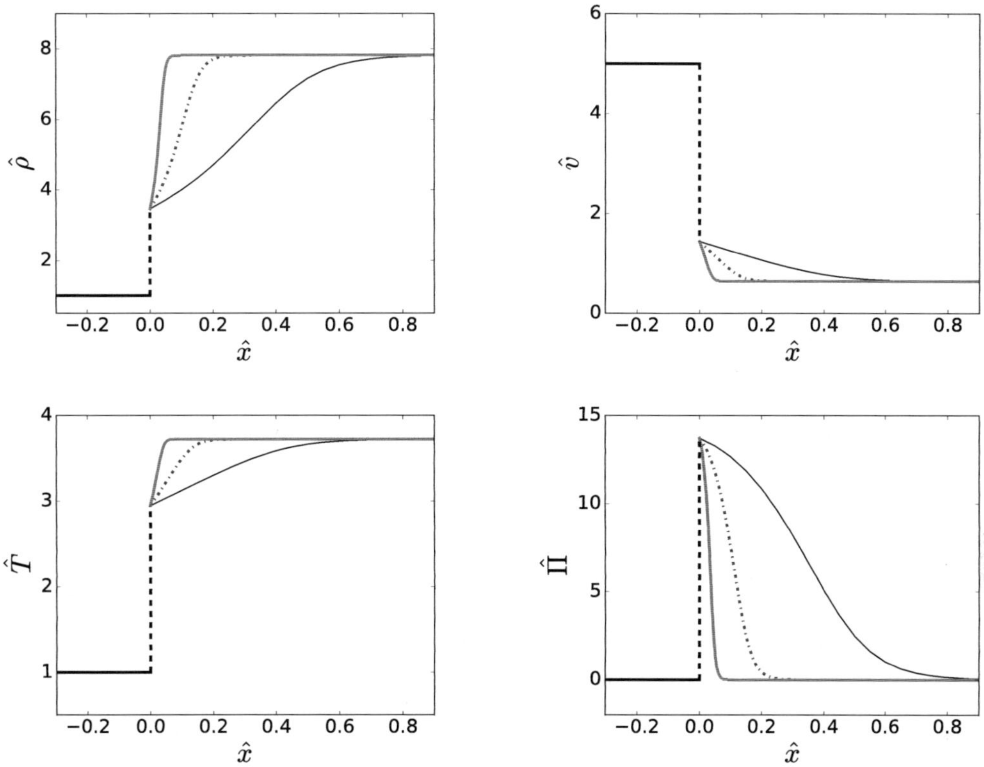

Figure 3. Profiles of the dimensionless mass density, velocity, temperature and dynamic pressure predicted by ET_6 for $M_0 = 5$. Three cases with $n = 1$ (black thin curves), $n = 0$ (blue dot-dashed curves) and $n = -1$ (red thick curves) are shown. The vertical dashed lines at $\hat{x} = 0$ represent the jumps of physical quantities across the sub-shock.

characteristic velocities propagating in the positive direction. In these cases, the determination of the characteristic value of the shock velocity for the sub-shock formation still remains as an open problem. The recent development on this issue including the peculiar feature of the sub-shock formation based on ET_{14} was reported in [40].

(III) Another ET theories for polyatomic gases have been proposed such as a refined version of ET_6 which is expected to be valid also for dense gases [41], ET_7 dealing with rotational and vibrational modes individually [42] and the ET theory for mixtures of polyatomic gases [43]. The analysis of the shock wave structure based on these ET theories is now in progress. The results will soon be reported elsewhere.

Acknowledgments
The results contained in the present paper have been partially presented in "The Meeting on Nonequilibrium Thermodynamics and Statistical Physics: From Rational Modeling to Its Applications" held in Fukuoka (Japan) 2017. The authors thank Prof. Kazuo Aoki for providing the original figure of his paper. This work was partially supported by JSPS KAKENHI Grant Numbers JP16K17555 (S. T.) JP15K21452 (T.A.) and by National Group of Mathematical Physics GNFM-INdAM (T. R.).

IOP Conf. Series: Journal of Physics: Conf. Series **1035** (2018) 012009 doi:10.1088/1742-6596/1035/1/012009

References

[1] Vincenti W G and Kruger C H Jr. 1965 *Introduction to Physical Gas Dynamics* (New York, London, Sydney: John Wiley and Sons)
[2] Zel'dovich Ya B and Raizer Yu P 2002 *Physics of Shock Waves and High-Temperature Hydrodynamic Phenomena* (Mineola, New York: Dover Publications)
[3] Smiley E F, Winkler E H and Slawsky Z I 1952 *J. Chem. Phys.* **20** 923
[4] Smiley E F and Winkler E H 1954 *J. Chem. Phys.* **22** 2018
[5] Griffith W C and Bleakney W 1954 *Am. J. Phys.* **22** 597
[6] Griffith W, Brickl D and Blackman V 1209 *Phys. Rev.* **102** 1209
[7] Johannesen N H, Zienkiewicz H K, Blythe P A and Gerrard J H 1962 *J. Fluid Mech.* **13** 213
[8] Griffith W C and Kenny A 1957 *J. Fluid Mech.* **3** 286
[9] Gilbarg D and Paolucci D 1953 *J. Rat. Mech. Anal.* **2** 617
[10] Bethe H A and Teller E 1941 Deviations from Thermal Equilibrium in Shock Waves (reprinted by Engineering Research Institute: University of Michigan)
[11] Meixner J 1943 *Ann. Physik* **43** 470
[12] Meixner J 1952 *Acoustica* **2** 101
[13] Müller I and Ruggeri T 1998 *Rational Extended Thermodynamics. 2nd edn.* (New York: Springer)
[14] Arima T, Taniguchi S, Ruggeri T and Sugiyama M 2012 *Cont. Mech. Thermodyn.* **24** 271
[15] Ruggeri T and Sugiyama M 2015 *Rational Extended Thermodynamics beyond the Monatomic Gas* (Cham, Heidelberg, NewYork, Dordrecht, London: Springer)
[16] Taniguchi S, Arima T, Ruggeri T and Sugiyama M 2014 *Phys. Rev. E* **89** 013025
[17] Taniguchi S, Arima T, Ruggeri T and Sugiyama M 2014 *Acta Appl. Math.* **132** 583
[18] Taniguchi S, Arima T, Ruggeri T and Sugiyama M 2014 *Phys. Fluids* **26** 016103
[19] Arima T, Taniguchi S, Ruggeri T and Sugiyama M 2012 *Phys. Lett. A* **376** 2799
[20] Arima T, Ruggeri T, Sugiyama M and Taniguchi, S 2014 *Meccanica* **49** 2181
[21] Arima T, Ruggeri T, Sugiyama M and Taniguchi S 2015 *International J. Non-Linear Mech.* **72** 6
[22] Taniguchi S, Arima T, Ruggeri T and Sugiyama M 2016 *International J. Non-Linear Mech.* **79** 66
[23] Bird G A 1994 *Molecular Gas Dynamics and the Direct Simulation of Gas Flows* (Oxford: Oxford Univ Press)
[24] Chapman S and Cowling T G 1991 *The Mathematical Theory of Non-Uniform Gases* (Cambridge: Cambridge Univ Press)
[25] Zhdanov V M 1968 *Soviet Phys. JETP* **26** 1187
[26] MacCormack F J 1968 *Phys. Fluids* **11** 2533
[27] MacCormack F J 1970 *Phys. Fluids* **13** 1446
[28] Mallinger F 1998 Generalization of the Grad theory to polyatomic gases (Research Report RR-3581, INRIA)
[29] Kosuge S, Aoki K and Goto T 2016 *AIP Conference Proceedings* **1789** 180004
[30] Kosuge S and Aoki K 2018 *Phys. Rev. Fluids* **3** 023401
[31] Arima T, Ruggeri T, Sugiyama M and Taniguchi S 2016 *Ricerche di Matematica* **65** 263
[32] Pavić M, Ruggeri T and Simić S 2013 *Physica A* **392** 1302
[33] Arima T, Mentrelli A and Ruggeri T 2014 *Annals of Physics* **345** 111
[34] Ruggeri T 2016 *Bulletin of the Institute of Mathematics Academia Sinica (New Series)* **11** No. 1, 1
[35] Bisi M, Ruggeri T and Spiga G 2018 *Kinetic and Related Models* **11** 71
[36] Boillat G and Ruggeri T 1998 *Cont. Mech. Thermodyn.* **10** 285
[37] Weiss W 1995 *Phys. Rev. E* **52** R5760
[38] 1983 *JSME Data Book: Thermophysical Properties of Fluids* (Tokyo: Japan Society of Mechanical Engineers)
[39] Cramer M S 2012 *Phys. Fluids* **24** 066102
[40] Taniguchi S and Ruggeri T 2018 *International J. Non-Linear Mech.* **99** 69
[41] Arima T, Ruggeri T and Sugiyama M 2017 *Phys. Rev. Fluids* **2** 013401
[42] Arima T, Ruggeri T and Sugiyama M 2017 *Phys. Rev. E* **96** 042143
[43] Arima T, Ruggeri T, Sugiyama M and Taniguchi S 2017 *Rend. Lincei Mat. Appl.* **28** 495

IOP Conf. Series: Journal of Physics: Conf. Series **1035** (2018) 012010 doi:10.1088/1742-6596/1035/1/012010

On the smoothing of point force in the two-way coupling simulation of polymer-laden turbulent flow

Takeshi Watanabe and Toshiyuki Gotoh
Department of physical science and engineering, Nagoya Institute of Technology, Japan
E-mail: watanabe@nitech.ac.jp

Abstract. Effects of smoothing of point force used in the two-way coupling simulation of particle-laden turbulent flow are examined for one dimensional simple model and for the Eulerian-Lagrangian simulations of polymers in turbulence. We show that the regularization by using Gaussian function yields preferable behavior in evaluating the smoothed point force compared to the conventional methods used in the previous studies. An efficient computational scheme to use the approximated Gaussian function is newly developed for the two-way coupling simulation in which the turbulent field is solved by using the spectral method. This scheme is applied to the Eulerian-Lagrangian simulations for examining polymer-turbulence interactions.

1. Introduction

It is very important to understand interaction between particles and turbulence because the issue is related to various fields in science and engineering [1]. One of the important problems is to understand the mechanism of growth of droplets in cloud [2], where the effect of turbulent mixing on coagulation and coalescence of droplets is crucial [3]. The other example is the drag reduction phenomenon observed in turbulent channel flow of non-Newtonian fluid where the long-chain polymers dispersed in the solvent fluid are stretched by strong local shear in turbulence, leading to the modification of turbulence properties [4, 5].

In many cases the particle size is comparable to or smaller than the dissipation length in turbulence. This suggests that a small particle is approximated by a mass point. When the Stokes drag force is dominant over the forces acting on point particle, the governing equations of motion for point particles (particle radius a; total particle number N_p) in fluid are given by [6].

$$\frac{d\boldsymbol{x}_p^i(t)}{dt} = \boldsymbol{v}_p^i(t), \quad m_p \frac{d\boldsymbol{v}_p^i(t)}{dt} = -\zeta \left[\boldsymbol{v}_p^i(t) - \boldsymbol{u}\left(\boldsymbol{x}_p^i(t), t\right)\right] + \boldsymbol{F}^i \quad (i = 1, \cdots, N_p), \tag{1}$$

where m_p is a particle mass and $\zeta = 6\pi\nu_f\rho_f a$ is Stokes drag coefficient, ν_f and ρ_f being the kinematic viscosity and density of solvent fluid, respectively. The term $\boldsymbol{F}^i$ represents the remaining forces like the repulsive force among particles, buoyancy force, etc., and $\boldsymbol{u}(\boldsymbol{x}, t)$ is the velocity field of the solvent fluid. The reaction force due to the suspended point particles acts on the incompressible fluid which obeys the Navier-Stokes (NS) equations. This can be expressed by the ensemble of point force as follows [7].

$$\frac{\partial \boldsymbol{u}}{\partial t} + \boldsymbol{u} \cdot \nabla \boldsymbol{u} = -\frac{1}{\rho_f}\nabla p + \nu_f \nabla^2 \boldsymbol{u} + \frac{1}{\rho_f}\sum_{i=1}^{N_p} \zeta \left[\boldsymbol{v}_p^i(t) - \boldsymbol{u}\left(\boldsymbol{x}_p^i(t), t\right)\right] \delta\left(\boldsymbol{x} - \boldsymbol{x}_p^i(t)\right), \tag{2}$$

IOP Conf. Series: Journal of Physics: Conf. Series **1035** (2018) 012010 doi:10.1088/1742-6596/1035/1/012010

with $\nabla \cdot \boldsymbol{u} = 0$, where $p(\boldsymbol{x}, t)$ is the pressure field and $\delta(\boldsymbol{x})$ is the Dirac delta function. To obtain the drag force acting on point particle numerically, fluid velocity at the particle position, $\boldsymbol{u}\left(\boldsymbol{x}_p^i(t), t\right)$, must be evaluated by interpolating the data of fluid velocity on the computational grid points. The evaluated point force is then returned to fluid by distributing it over the surrounded grid points because we cannot deal with $\delta(\boldsymbol{x})$ in computation. This calculation is called the two-way coupling.

In the past many studies have conducted to clarify the nature of interaction between the ensemble of point particles and turbulence by performing two-way coupling simulations. The modification of the kinetic energy spectrum of isotropic turbulence due to the presence of inertial particles was discussed in [8], where it was observed that the spectral excitation at high wavenumber grows with increasing the Stokes number $S_t = m_p/(\zeta \tau_K)$, where τ_K is the characteristic time scale in turbulence at small scale, suggesting strong modification of dissipation scale dynamics in turbulence. Our previous studies [9, 10] investigated the effect of polymer additives on isotropic turbulence by performing the Eulerian-Lagrangian simulations, where the long-chain polymer molecule is represented by the beads-spring model. Although these analysis can be performed with reasonable computational cost even when the number of particles becomes much larger, there remains two kinds of problem in two-way coupling calculation: 1) we do not have enough knowledge about interactions between turbulence and ensemble of particles because the flow around particles is not resolved on the computational grid points, 2) the results obtained by two-way coupling calculation depend on interpolation/distribution methods representing the interaction between particles and turbulence.

The purpose of the present study is to clarify the second problem described above. That is, the question is how the details of smoothing method for point force in two-way coupling calculation affect the interaction between the ensemble of particles and turbulent flow. First we consider one dimensional model with three kinds of test function for which the analytical solution is given. This analysis brings insights into the validity/limitation of the method. Then we examine the two-way coupling simulations of polymer-laden isotropic turbulence.

This paper is organized as follows. In Section 2 we investigate three smoothing methods for the test model of point force. The efficiency and accuracy of the methods are discussed in detail. In Section 3 the two-way coupling simulations for polymer-laden isotropic turbulence are performed and examined from the viewpoint same as in Section 2. We present conclusions obtained in the present study in Section 4.

2. Characteristics of smoothed point force: case of the simple model

The problem when performing the two-way coupling simulations of equations (1) and (2) is how the delta function $\delta(\boldsymbol{x} - \boldsymbol{x}_p^i)$ in point force is approximated. From the computational point of view, the sufficient accuracy and reasonable computational cost are simultaneously required when evaluating the point forces with $N_p \gg 1$ case. In this section we investigate this problem by using the simple model.

2.1. Consideration by one-dimensional test problem

In this subsection we discuss the one-dimensional problem represented by

$$T(x) = \frac{L}{N_p} \sum_{i=1}^{N_p} t_i \delta(x - x_i), \tag{3}$$

where x_i $(i = 1, \cdots, N_p)$ indicates the position of particle which locates within $-L/2 \leq x \leq L/2$, and t_i is a physical quantity characterizing a particle at x_i. When we evaluate equation (3) numerically, $T(x)$ is discretized as $T(x_j)$ with $x_j = j\Delta x$ $(j = -N/2, \cdots, N/2)$, Δx is the grid spacing, so that the delta function $\delta(x)$ must be replaced by some approximated function $\delta_\Delta(x)$ whose support is in the range $|x| \leq \Delta$. Here we choose $\Delta = \Delta x$. We rewrite equation (3) as

$$T_\Delta(x) = \frac{L}{N_p}\sum_{i=1}^{N_p} t_i \delta_\Delta(x - x_i), \tag{4}$$

where

$$\int_{-\infty}^{\infty} \delta_\Delta(x)\, dx = 1, \quad \lim_{\Delta\to 0} \delta_\Delta(x) = \delta(x) \tag{5}$$

must be satisfied. Computational cost when computing (4) comes from 1) evaluating $\delta_\Delta(x-x_i)$, 2) product of t_i and $\delta_\Delta(x - x_i)$, and 3) accessing the memory space that the data set of $T(x_j)$ is maintained (array variable for $T(x_j)$). We write the number of times of abovementioned operations as C_1,C_2 and C_3, respectively. In this study we consider the following three cases of $\delta_\Delta(x)$.

i) *Rectangular function* : this is defined by

$$\delta^r_\Delta(x) = \begin{cases} \dfrac{1}{\Delta} & (|x| < \Delta/2), \\ 0 & (|x| \geq \Delta/2). \end{cases} \tag{6}$$

This may be the simplest case for calculation. The evaluation of $T(x_j)$ means t_i/Δ is put on a grid point that $|x_i - x_j|$ is minimum. In this case, the number of times of the operation is estimated as $C_1 = 0$, $C_2 = 1$ and $C_3 = N_p$.

ii) *Triangular function*: this is defined by

$$\delta^t_\Delta(x) = \begin{cases} \dfrac{1}{\Delta}\left(1 - \dfrac{|x|}{\Delta}\right) & (|x| < \Delta), \\ 0 & (|x| \geq \Delta). \end{cases} \tag{7}$$

Using this function when computing (4) means that t_i with multiplying the weight determined by $\delta^t_\Delta(x_i - x_j)$ or $\delta^t_\Delta(x_{j+1} - x_i)$ is distributed on grid points of x_j and x_{j+1} which satisfy $x_j < x_i < x_{j+1}$. This also corresponds to the method using the linear interpolation scheme in which the interaction between particle and fluid occurs within the surrounded grid points [7, 9]. The number of times of operation when computing equation (7) is $C_1 = 2N_p$ which is mostly dominated by evaluating the distance $|x_i - x_j|$. While the number of times of the product in (4) is $C_2 = 2N_p$, so that the number of times of accessing the memory for $T(x_j)$ is $C_3 = 2N_p$.

iii) *Gaussian function*: this is defined by

$$\delta^g_\Delta(x) = \frac{1}{\sqrt{2\pi}\Delta}\exp\left(-\frac{x^2}{2\Delta^2}\right). \tag{8}$$

Recently the rigorous form of regularized point force was derived from analyzing the unsteady Stokes equations, in which the regularized point force involves Gaussian function for $\delta_\Delta(x)$ [11]. Because the support of Gaussian function is almost in the range $-3\Delta \leq x \leq 3\Delta$, we evaluate the values of $\delta_\Delta(x)$ on 7 grid points per particle. In this case we have the estimations $C_1 = 7N_p$,$C_2 = 7N_p$ and $C_3 = 7N_p$. Note that here C_1 is the number of times of the calculation for $\exp\left(-x^2/(2\Delta^2)\right)$ whose calculating cost is much higher than the case of $\delta^t_\Delta(x)$.

To test the effect of variation of $\delta_\Delta(x)$ on resulting $T_\Delta(x)$, we examine the test problem that t_i in equation (4) is given by $t_i = \sin x_i$. $T_\Delta(x)$ is defined in $-\pi \leq x \leq \pi$, and supposed to have the periodicity $L = 2\pi$. The computational domain is discretized using N grid points as

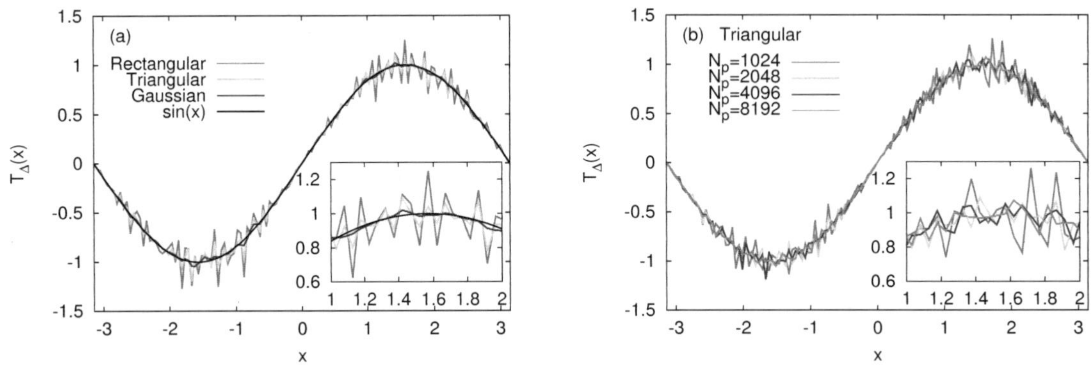

Figure 1. Comparison of $T_\Delta(x)$ computed by equation (4). (a) For three cases of $\delta_\Delta(x)$ when $N_p = 2048$, (b) for various N_p in the case of the triangular function. In each figure, the inset shows the enlarged plot of the interval $1 \le x \le 2$.

$x_j = j\Delta x$ ($j = -N/2.\cdots, N/2-1$), where the grid spacing is $\Delta x = 2\pi/N$. We fix $N = 128$, while the number of particles N_p is varied as $N_p = 1024, 2048, 4096$ and 8192. In this case the number of particles within an unit width Δx is $N_p/N = 8, 16, 32$ and 64. The position of particles x_i is determined by using the uniform random number in the interval $[-\pi, \pi]$. We compute equation (4) for above three cases of $\delta_\Delta(x)$ and compare the obtained results with the analytical result of $\sin x$ which is given in the limit $N_p \to \infty$.

Figure 1 (a) shows the comparison of obtained $T_\Delta(x)$ for the cases of i)-iii) when $N_p = 2048$. Zig-zag fluctuations around the exact sine curve are observed when the rectangular and triangular functions are used, while the curve obtained by using the Gaussian function is smooth in the whole region and close to the exact one. Figure 1 (b) indicates how $T_\Delta(x)$ depends on N_p in the case for the triangular function. As expected, the amplitude of zig-zag fluctuations decreases with increasing N_p, although small fluctuations around the curve of $\sin x$ remain for $N_p = 8192$ ($N_p/N = 64$).

To quantify the deviation of resulting $T_\Delta(x)$ curves from the analytical result, we evaluate the L^2 norm

$$\sigma^2 = \frac{1}{L}\int_{-L/2}^{L/2} \left[T_\Delta(x) - \sin x\right]^2 \, dx \tag{9}$$

and examine its N_p-dependence. Figure 2 shows σ against N_p for above three cases of $\delta_\Delta(x)$. Result obtained by using Gaussian function is about one decade smaller than those of rectangular and triangular cases, indicating that the smoothing by Gaussian function leads to the better behavior when compared to that by other two cases. It is worth noting that σ decays with N_p in the power-law form $N_p^{-0.4}$ irrespective of the functional form of $\delta_\Delta(x)$. This slow decay suggests that much larger number of N_p is required to obtain well converged $T_\Delta(x)$, especially the case of rectangular function is serious.

2.2. Efficient calculation of smoothed point force using Gaussian function

In the previous subsection, we obtained that the smoothed point force using Gaussian function yields the accurate behavior of the resulting function which is close to the analytical result. However the computational cost to obtain equation (4) when using $\delta^g_\Delta(x)$ was much higher than other methods. Moreover if we consider the problem in three dimension, for example, the number of times of evaluating $\delta^g_\Delta(\boldsymbol{x})$ is increased by $C_1 = 7^3 N_p$, which is considerable computational cost when $N_p \gg N$ case like the polymer-laden flows is conducted [9]. In this subsection, we newly

IOP Conf. Series: Journal of Physics: Conf. Series **1035** (2018) 012010 doi:10.1088/1742-6596/1035/1/012010

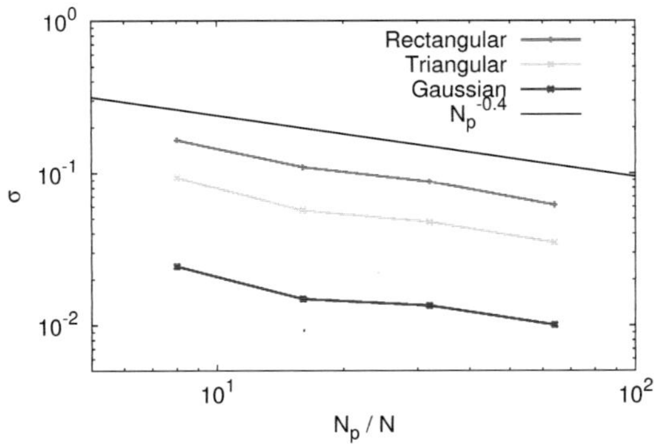

Figure 2. Degree of the deviation from the analytical form, represented by σ (equation (9)), against the particle number N_p for three cases of $\delta_\Delta(x)$.

propose the approximated but efficient computational method of equation (4) with making use of the Gaussian function when the turbulent motion is solved by Fourier spectral method.

It is interesting to note that the relationship between $T(x)$ and $T_\Delta(x)$ is expressed by

$$\hat{T}_\Delta(k) = \hat{\delta}_\Delta(k)\hat{T}(k) \tag{10}$$

in Fourier space, where the variable with symbol of top hat $\hat{g}(k)$ represents the Fourier amplitude of $g(x)$. The rectangular (6) and Gaussian (8) functions are respectively represented in Fourier space by

$$\hat{\delta}^r_\Delta(k) = \frac{\sin(k\Delta/2)}{k\Delta/2}, \quad \hat{\delta}^g_\Delta(k) = \exp\left[-\frac{1}{2}(k\Delta)^2\right]. \tag{11}$$

The idea is to find an appropriate function $\hat{d}_\Delta(k)$ leading to the relationship

$$\hat{\delta}^g_\Delta(k) \simeq \hat{d}_\Delta(k)\hat{\delta}^r_\Delta(k) \tag{12}$$

because

$$\hat{T}^g_\Delta(k) \simeq \hat{d}_\Delta(k)\hat{T}^r_\Delta(k), \tag{13}$$

where $\hat{T}^g_\Delta(k)$ and $\hat{T}^r_\Delta(k)$ are evaluated by using the Gaussian and rectangular functions for $\delta_\Delta(x)$, respectively. Once we know the function $\hat{d}_\Delta(k)$, the procedure to compute the approximated $T^g_\Delta(x)$ is as follows.

1) Evaluating $T^r_\Delta(x)$ in physical space using equation (4), and transforming it into the Fourier amplitude $\hat{T}^r_\Delta(k)$ with the aid of the fast Fourier transform (FFT) algorithm.
2) Evaluating $\hat{d}_\Delta(k)\hat{T}^r_\Delta(k)$ in Fourier space and transforming it into the physical space variable.

If $N_p \gg N$, most of the computational costs come from accessing an array for $T_\Delta(x_j)$, indicating that the computational cost for the above scheme is comparable with that of evaluating $T^r_\Delta(x)$. The left problem is how $\hat{d}_\Delta(k)$ is determined. We propose the following Gaussian function as

$$\hat{d}_\Delta(k) = \exp\left[-\frac{1}{2}(k\gamma)^2\right], \quad \gamma = \sqrt{\frac{11}{12}}\Delta. \tag{14}$$

The factor $\sqrt{11/12}$ originates from the condition that the first and second order terms of the Taylor series expansion of $\hat{d}_\Delta(k)\hat{\delta}^r_\Delta(k)$ with (14) are equalized to those of $\hat{\delta}^g_\Delta(k)$ in (11). To check the relationship (12) with (14), we compare the approximated form of $\hat{\delta}^g_\Delta(k)$ with exact one. Figure 3 (a) indicates the resulting curves when $\Delta = \pi/64$. Approximated form evaluated

IOP Conf. Series: Journal of Physics: Conf. Series **1035** (2018) 012010 doi:10.1088/1742-6596/1035/1/012010

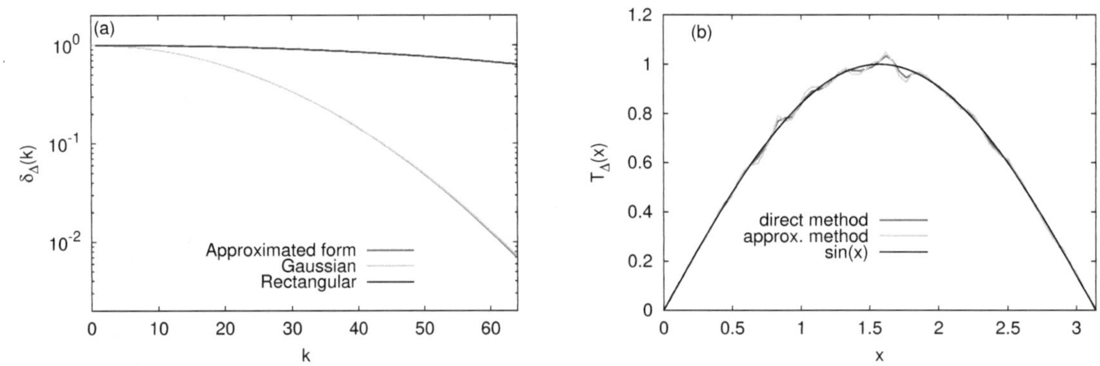

Figure 3. (a) Validation of the proposed form equation (12) with equation (14) (approximated Gaussian function). (b) Comparison of $T_\Delta(x)$ computed by the direct method equation (4) with that by the approximated method using equations (12) and (14), where the resulting curves in the range $0 \le x \le \pi$ are shown for simplicity.

by equation (12) with (14) almost coincides with the result by exact Gaussian function as far as $k \le 64$. Figure 3 (b) indicates $T_\Delta(x)$ evaluated by equation (4) with Gaussian function and by the approximated method we proposed here when $N_p = 8192$. It is confirmed that both curves are almost similar, representing that the approximated method can reproduce the result by the direct method using (4) with much lower computational cost.

3. Two-way coupling simulation for polymer-laden isotropic turbulence

In this section we discuss how the details of smoothing of the point force affect the nature of interactions between polymer and turbulent flow by performing the two-way coupling simulations.

3.1. Fundamental equations of motion

We briefly review the fundamental equations of motion. The motion of long-chain polymer in fluid is described by the beads-spring model [12]. In this study we consider the simplest case that the elastic nature of polymer chain is represented by 2-beads connected by nonlinear spring, called the dumbbell model [13]. This is given by

$$\frac{d\boldsymbol{R}^{(n)}}{dt} = \boldsymbol{u}_1^{(n)} - \boldsymbol{u}_2^{(n)} - \frac{1}{2\tau} f\left(\frac{|\boldsymbol{R}^{(n)}|}{L_{max}}\right)\boldsymbol{R}^{(n)} + \frac{r_{eq}}{\sqrt{2\tau}}\left(\boldsymbol{W}_1^{(n)} - \boldsymbol{W}_2^{(n)}\right), \tag{15}$$

$$\frac{d\boldsymbol{r}_g^{(n)}}{dt} = \frac{1}{2}\left(\boldsymbol{u}_1^{(n)} + \boldsymbol{u}_2^{(n)}\right) + \frac{r_{eq}}{\sqrt{8\tau}}\left(\boldsymbol{W}_1^{(n)} + \boldsymbol{W}_2^{(n)}\right), \quad \boldsymbol{u}_\alpha^{(n)} \equiv \boldsymbol{u}(\boldsymbol{x}_\alpha^{(n)}(t), t), \tag{16}$$

where $\boldsymbol{R}^{(n)}(t)$ and $\boldsymbol{r}_g^{(n)}(t)$ are, respectively, the end-to-end vector and the center-of-mass vector of the n-th dumbbell. We adopt the finitely extensible nonlinear elastic (FENE) model $f(z) = 1/(1 - z^2)$ for the elastic force of a dumbbell. In equation (15), L_{max} is the maximum extension length of the dumbbell because $f(z \to 1) = \infty$. The term $\boldsymbol{W}_{1,2}^{(n)}(t)$ indicates a random force representing the Brownian motion of particles in the solvent fluid, which obeys Gaussian statistics with a white-in-time correlation of $\langle W_{\alpha,i}^{(n)}(t)\rangle = 0$ and $\langle W_{\alpha,i}^{(m)}(t) W_{\beta,j}^{(n)}(s)\rangle = \delta_{\alpha\beta}\delta_{ij}\delta_{mn}\delta(t-s)$, where $\langle\cdots\rangle$ denotes the ensemble average. The subscripts α, β, i, j, n, and m take the values $(\alpha, \beta) = 1$ or 2, $(i, j) = 1$, 2, and 3, and $(n, m) = 1, 2, \cdots, N_p$, respectively. Moreover, δ_{ij} denotes the Kronecker delta. The constants $\tau \equiv \zeta/(4k)$ and $r_{eq} \equiv \sqrt{k_B T/k}$ are, respectively, the relaxation time and the equilibrium length of the dumbbell under $\boldsymbol{u}(\boldsymbol{x}, t) = \boldsymbol{0}$. Here, k is the spring constant, k_B and T are the Boltzmann constant and temperature, respectively.

The turbulent velocity field obeys the continuity equation $\nabla \cdot \boldsymbol{u} = 0$ for an incompressible fluid and the NS equations

$$\frac{\partial \boldsymbol{u}}{\partial t} + \boldsymbol{u} \cdot \nabla \boldsymbol{u} = -\nabla p + \nu_f \nabla^2 \boldsymbol{u} + \nabla \cdot \boldsymbol{T}^p. \tag{17}$$

Here, ρ_f is set to unity and is equal to the density of bead ρ_p representing the polymer. $\boldsymbol{T}^p(\boldsymbol{x}, t)$ is the polymer stress tensor due to the force acting on the fluid from the dispersed dumbbells and is defined by

$$T^p_{ij}(\boldsymbol{x}, t) = \frac{\nu_f \eta}{\tau} \left(\frac{L^3_{box}}{N_p} \right) \sum_{n=1}^{N_p} \left[\frac{R_i^{(n)} R_j^{(n)}}{r^2_{eq}} f\left(\frac{|\boldsymbol{R}^{(n)}|}{L_{max}} \right) - \delta_{ij} \right] \delta(\boldsymbol{x} - \boldsymbol{r}_g^{(n)}), \tag{18}$$

where N_p is the total number of dumbbells used in simulations. η is the ratio of the zero-shear viscosity ν_p of the polymer to the solution viscosity ν_f $(\eta = \nu_p/\nu_f)$, and is the parameter to control the concentration of polymer solution.

3.2. Summary of numerical setup

The two-way coupling simulations of equations (15) through (18) are performed in a periodic box with periodicity $L_{box} = 2\pi$ using the pseudo-spectral method in space and the second-order Runge–Kutta method in time. A total of 128^3 grid points are set to solve the NS equations. We consider the case of decaying isotropic turbulence, as discussed in our previous studies [9, 10]. The initial velocity field is given by the random solenoidal field obeying Gaussian statistics and yields an energy spectrum $E(k, 0) = 16\sqrt{2/\pi}\,(u_0^2/k_0)\,(k/k_0)^4 \exp\left(-2\,(k/k_0)^2\right)$ with constants $u_0 = 1$ and $k_0 = 2$.

The total number of dumbbells in computation is $N_p = 5 \times 10^8$. The dumbbells are uniformly and randomly distributed over the computational domain. The initial configuration of each dumbbell is set according to $\boldsymbol{R}^{(m)}(0) = \sqrt{3} r_{eq} \boldsymbol{n}^{(m)}$, where $\boldsymbol{n}^{(m)}$ is an unit vector whose orientation is randomly distributed. We fix $\eta = 0.1$ and $\tau = 20$ whose values correspond to those used for Run D3 in [10].

The computation is parallelized by writing the code using Message Passing Interface (MPI) and Open MP. For details about parallel computation, please refer to the reference [14]. We examine the three cases of smoothing method discussed in the previous section. The approximated method for making use of the Gaussian function, proposed in the previous section, is introduced in the present analysis. For comparison we present the results obtained by our previous scheme [10] (named by WG14), in which we evaluate the polymer stress on the regular and staggered grid points, respectively, with making use of the triangular function and by taking the average of them in the Fourier space. For more details please see appendix in the reference [10].

3.3. Results

Figure 4 (a) shows the temporal evolutions of the kinetic energy spectrum which is defined by

$$E(k, t) = \sum_{\boldsymbol{k}}{}' \frac{1}{2} |\hat{\boldsymbol{u}}(\boldsymbol{k}, t)|^2 \tag{19}$$

IOP Conf. Series: Journal of Physics: Conf. Series **1035** (2018) 012010 doi:10.1088/1742-6596/1035/1/012010

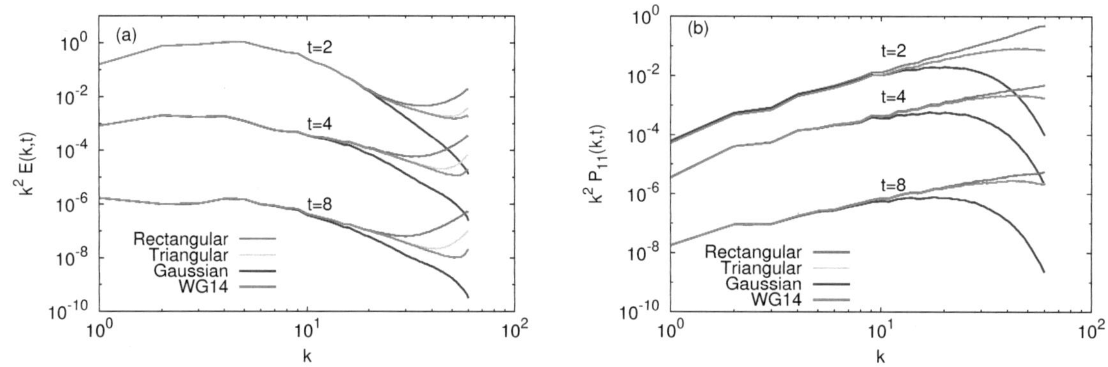

Figure 4. Temporal evolutions of (a) the kinetic energy and (b) variance of polymer stress ($T_{11}^{(p)}$) spectra evaluated for the four cases of the smoothing method when computing $T_{ij}^{(p)}$ (equation (18)). All spectra are plotted by multiplying k^2 to stress the behavior in high wavenumber range, and the curves obtained at $t = 4$ and $t = 8$ are shifted lower for being easy to see.

in terms of the Fourier amplitude $\hat{\boldsymbol{u}}(\boldsymbol{k}, t)$ where $\sum'_{\boldsymbol{k}}$ is the summation taken over the spherical shell within $k - \Delta k/2 < |\boldsymbol{k}| \leq k + \Delta k/2$ in the wavenumber space. In this figure $k^2E(k,t)$ is plotted to stress the behavior in the high wavenumber range. It is seen from this figure that curves of $E(k,t)$ in the lower wavenumber range collapse well irrespective of the choice of $\delta_\Delta(x)$, indicating that large-scale motion of turbulence is not affected by the details of smoothing method. However there are significant difference of the behavior of $k^2E(k,t)$ in the high wavenumber range for different $\delta_\Delta(\boldsymbol{x})$. The worst case is to use the rectangular function because the spectra near the cutoff wavenumber pile up when compared to other cases. This means that the velocity gradient fluctuations like the local energy dissipation are not well resolved on grids. On the other hand the results with the Gaussian function show the rapid decay of spectrum in its spectral tail, suggesting the smooth behavior of fluctuating velocity gradient field. This trend is also directly observed in the spectral behavior of the polymer stress tensor shown in figure 4 (b), where we indicate the spectrum of $T_{11}^p(\boldsymbol{x}, t)$, $P_{11}(k) = \sum'_{\boldsymbol{k}} \frac{1}{2}|\hat{T}_{11}^p(\boldsymbol{k}, t)|^2$, with multiplying k^2. Note that the results obtained by using the triangular function almost coincide with those by WG14 because the smoothing method in WG14 is based on the triangular function. Difference among them appears for the intermittent fluctuations of polymer stress $\boldsymbol{T}^p$ in physical space (see figure 12 (b) in reference [10]).

As shown in figure 4 (a), $k^2E(k,t)$ obtained by using the smoothed point force with Gaussian function is steeper than that obtained by our previous method (WG14) [10] in which the power-law decay of the form $E(k,t) \propto k^{-\alpha}$ with $\alpha = 4 \sim 5$ is observed in the final period of decaying turbulence. Because the support of Gaussian function in physical space is wider than that of rectangular and triangular functions, it is plausible that the amplitude of a spectrum evaluated by the smoothed point force with Gaussian function damps strongly in the high wavenumber range more than that evaluated by using other functions. This fact suggests if the Fourier modes are excited in the wide range of scale like the power-law decay of $E(k,t)$, we need great care for the effect of smoothing with a Gaussian function on the resulting spectral behavior. To examine the impact for using $\delta_\Delta^g(\boldsymbol{x})$ on spectral behavior in the low wavenumber range, we perform an additional simulation which has higher resolution with grid points 256^3 but other physical conditions are unchanged. Note that because Δx with $N = 256^3$ is half of that with $N = 128^3$, the width Δ of Gaussian function is also half of that with the case of 128^3. Figure 5 shows the comparison of resulting curves of $k^2E(k,t)$. Although $E(k,t)$ in lower wavenumber

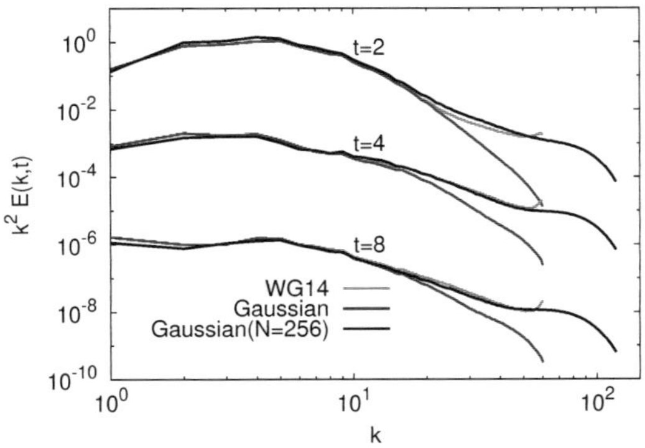

Figure 5. Comparison of the behavior of kinetic energy spectra for both the results using $N = 128^3$ grid points and $N = 256^3$ ones with using the Gaussian function. All spectra are plotted by multiplying k^2 to stress the behavior in high wavenumber range, and the curves obtained at $t = 4$ and $t = 8$ are shifted lower for being easy to see.

range almost coincides with each other, we observe different behavior at the high wavenumbers where the curves obtained by 256^3 grid points are shallower than those by $N = 128^3$ and follow the results of our previous method [10]. This result suggests that we must pay attention to the spatial resolution when using the Gaussian function and examining the physical process in the high wavenumber range.

Finally we comment on the computational time required for simulating the present polymer-laden isotropic turbulence. The real computational time required for time development per one time step Δt is summarized in table 1. As expected the computational time required for the simulation using the rectangular function is shortest throughout the present runs, and this time is almost comparable to that using the approximated Gaussian function. The computational time of our previous method [10] is about twice of that using Gaussian function, indicating the efficiency of computation using approximated Gaussian function developed in the present study.

4. Conclusion

In this paper we discussed how the details of smoothing method for point force of two-way coupling simulations in polymer-laden turbulent flow affect the turbulence statistics. We considered the three types of functions (rectangular, triangular and Gaussian) to approximate the delta function when evaluating the point force. First we examined the accuracy of the methods by applying to the simple one-dimensional problem for which the analytical form is known. It was found that the rectangular and triangular functions showed zig-zag fluctuations of the resulting curves. On the other hand the results obtained by using the Gaussian function showed the accurate behavior close to the analytical result, suggesting the advantage to use it for two-way coupling problem. We also proposed the efficient computational method to use the Gaussian function when smoothing the point force, and confirmed the validity of our idea. This method was applied to the two-way coupling simulations of polymer-laden isotropic turbulence, where we could reproduce the results obtained by the method in [10] with satisfying accuracy

Table 1. Comparison of mean computational time required for time development par one time step Δt. "Gaussian" in table means to use the approximated form (12) with (14). Note that Δt of the computation using $N = 256^3$ grid points is half of that of $N = 128^3$ case.

	Rectangular	Triangular	Gaussian	WG14	Gaussian ($N = 256^3$)
time (s)	1.85	4.95	2.18	4.65	5.15

IOP Conf. Series: Journal of Physics: Conf. Series **1035** (2018) 012010 doi:10.1088/1742-6596/1035/1/012010

and better efficiency of computation.

Acknowledgments

The computational support provided by Japan High Performance Computing and Networking, Large-scale Data Analyzing and Information Systems (JHPCN) (jh170013), by NIFS Collaboration Research program (NIFS16KNXN313,NIFS18KNXN366), and by Collaborative Research Project on Computer Science with High-Performance Computing in Nagoya University are gratefully acknowledged. The authors were supported by JSPS KAKENHI Grant Numbers 26420106, 18K03925 and MEXT KAKENHI Grant Number 15H02218.

References

[1] Falkovich G, Gawedzki K and Vergassola M 2001 *Rev. Mod. Phys.* **73** 913
[2] Shaw R A 2003 *Annu. Rev. Fluid Mech.* **35** 183
[3] Saito I and Gotoh T 2018 *New J. Phys.* **20** 023001
[4] Lumley J L 1973 *J. Polymer Sci.: Macromolecular Reviews* **7** 263
[5] Proccacia I, L'vov V S and Benzi R 2008 *Rev. Mod. Phys.* **80** 225
[6] Maxey M R and Riley J J 1983 *Phys. Fluids* **26** 883
[7] Prosperetti A and Tryggvason G 2007 *Computational Methods for Multiphase Flow* (Cambridge: Cambridge University Press)
[8] Sundaram S and Collins L R 1999 *J. Fluid. Mech.* **379** 105
[9] Watanabe T and Gotoh T 2013 *J. Fluid. Mech.* **717** 535
[10] Watanabe T and Gotoh T 2014 *Phys. Fluids* **26** 035110
[11] Gualtieri, Picano F, Sardina G and Casciola C M 2015 *J. Fluid. Mech.* **773** 520
[12] Doi M and Edwards S F 1986 *The Theory of Polymer Dynamics* (New York: Oxford Univ. Press)
[13] Bird R B, Curtiss C F, Amstrong R C and Hassager O 1987 *Dynamics of Polymetric Liquids, Vol.2 Kinetic Theory, 2nd ed.* (New York: Wiley)
[14] Watanabe T and Gotoh T 2013 *J. Phys.: Conf. Ser.* **454** 012007

Institute of Physics
Dirac House, Temple Back
Bristol BS1 6BE UK

ISSN: 1742-6588
ISBN 978-1-5108-6410-8